James S. Fritz, Douglas T. Gjerde

Ion Chromatography

Further Reading

Journal of High Resolution Chromatography
ISSN 0935-6304 (12 issues per year)

Electrophoresis
ISSN 0173-0835 (18 issues per year)

Reference Work

J. Weiss
Ion Chromatography
2nd edition, 1995. ISBN 3-527-28698-5

James S. Fritz, Douglas T. Gjerde

Ion Chromatography

Third, completely revised and enlarged edition

Weinheim · New York · Chichester · Brisbane · Singapore · Toronto

Prof. Dr. James S. Fritz
Ames Laboratory
Iowa State University
332 Wilhelm Hall
Ames, IA 50010
USA

Dr. Douglas T. Gjerde
Transgenomic, Inc.
2032 Concourse Drive
San Jose, CA 95131
USA

Library of Congress Card No. applied for

A catalogue record for this book is available from the British Library

Die Deutsche Bibliothek – CIP Cataloguing-in-Publication-Data
A catalogue record for this publication is available from Die Deutsche Bibliothek

© WILEY-VCH Verlag GmbH, D-69469 Weinheim (Federal Republic of Germany), 2000

Printed on acid-free and chlorine-free paper

Composition: Kühn & Weyh, D-79111 Freiburg
Printing: Strauss Offsetdruck, D-69509 Mörlenbach
Bookbinding: J. Schäffer GmbH, D-67269 Grünstadt

Printed in the Federal Republic of Germany

Preface

Much has happened since the first edition appeared in 1982 and the second edition appeared in 1987. Ion chromatography has undergone impressive technical developments and has attracted an ever-growing number of users. The instrumentation has improved and the wealth of information available to the user has increased dramatically. Research papers and posters on new methodology and on applications in the power and semiconductor industries, pharmaceutical, clinical and biochemical applications and virtually every area continue to appear. An increasing number of papers on ion analysis by capillary electrophoresis is also included. Ion chromatography is now truly international in its scope and flavor.

This third edition is essentially an entirely new book. Our goal has been to describe the materials, principles and methods of ion chromatography in a clear, concise style. Whenever possible the consequences of varying experimental conditions have been considered. For example, the effects of the polymer structure and the chemical structure of ion-exchange groups and the physical form of the ion-exchange group attachment on resin selectivity and performance are discussed in Chapter 3.

Because commercial products are constantly changing and improving, the equipment used in ion chromatography is described in a somewhat general manner. Our approach to the literature of IC has been selective rather than comprehensive. Key references are given together with the title so that the general nature of the reference will be apparent. Our goal is to explain fundamentals, but also provide information in the form of figures and tables that can be used for problem solving by advanced users.

As well as covering the more or less "standard" aspects of ion chromatography, this is meant to be something of an "idea" book. The basic simplicity of ion chromatography makes it fairly easy to devise and try out new methods. Sometimes a fresh approach will provide the best answer to an analytical problem.

James S. Fritz, Ames, IA   November 1999
Douglas T. Gjerde, San Jose, CA

Acknowledgements

We would like to extend special acknowledgement for the support of our respective families for they, more than anything else, make life enjoyable and worthwhile.

We have received valuable help from a number of sources in writing this book. Ruthann Kiser (Dionex), Raaidah Saari-Nordhaus (Alltech), Dan Lee (Hamilton), Shree Karmarkar (Zellweger Analytics), and Helwig Schafer (Metrohm) have generously supplied various figures and other information. Also thanks to Y. S. Fung and Lau Kap Man (University of Hong Kong), Andy Zemann (Innsbruck University), and Dennis Johnson (Iowa State University), and former ISU students Greg Sevenich, Bob Barron, Youchun Shi, Weiliang Ding and Jie Li for various tables and figures. We thank Marilyn Kniss and Tiffany Nguyen for their hard work in preparing this manuscript to be sent to the publisher. We also thank Jeffrey Russell for his help in preparing the cover design.

The year 1999 marks the retirement from university teaching for one of us (JS). In fact, DG had the pleasure and honor of helping present the last university lecture of JS. This by no means marks the end of contributions to scientific discovery, and teaching made by JS. This will go on with new projects, publications, and correspondence. Nevertheless, DG would like to acknowledge the outstanding scientific accomplishments of JS that have been made through the years in ion chromatography and many other areas of analytical chemistry. DG would also like to wish JS many more years of fruitful and successful work.

James S. Fritz   Douglas T. Gjerde
Ames, Iowa   San Jose, California

Table of Contents

6   Anion Chromatography

11   Chemical Speciation

12   Method Development

1 Introduction and Overview

1.1 Introduction

The name "ion chromatography" applies to any modern method for chromatographic separation of ions. Normally, such separations are performed on a column packed with a solid ion-exchange material. But if we define chromatography broadly as a process in which separation occurs by differences in migration, capillary electrophoresis may also be included.

Ion chromatography is considered to be an indispensable tool in a modern analytical laboratory. Complex mixtures of anions or cations can usually be separated and quantitative amounts of the individual ions measured in a relatively short time. Higher concentrations of sample ions may require some dilution of the sample before introduction into the ion-chromatographic instrument. "Dilute and shoot" is the motto of many analytical chemists. However, ion chromatography is also a superb way to determine ions present at concentrations down to at least the low part per billion (μg/L) range. Although the majority of ion-chromatographic applications have been concerned with inorganic and relatively small organic ions, larger organic anions and cations may be determined as well.

Modern ion chromatography is built on the solid foundation created by many years of work in classical ion-exchange chromatography (see Chapter 2). The relationship between the older ion-exchange chromatography and modern ion chromatography is similar to that between the original liquid chromatography and the later high-performance liquid chromatography (HPLC) in which automatic detectors are used and the efficiency of the separations has been drastically improved. Ion chromatography as currently practiced is certainly "high performance" even though these words are not yet part of its name. Sometime in the future an even better form of ion chromatography (IC) may be dubbed HPIC.

1.2 Historical Development

Columns of ion-exchange resins have been used for many years to separate certain cations and anions from one another. Cations are separated on a cation exchange resin column, and anions are separated on a column containing an anion exchange resin. The most used types are as follows:

Polystyrene–⟨◯⟩–$SO_3^-H^+$ Polystyrene–⟨◯⟩–CH_2N^+, A^-

    Catex   Anex

For example, Na^+ and K^+ can be separated on a cation-exchange resin (Catex) column with a dilute solution of a strong acid (H^+) as the eluent (mobile phase). Introduction of the sample causes Na^+ and K^+ to be taken up in a band (zone) near the top of the column by ion exchange.

$$Resin–SO_3^-H^+ + Na^+, K^+ \rightleftarrows Resin–SO_3^-Na^+, K^+ + H^+$$

Continued elution of the column with an acidic eluent (H^+) introduces competition of H^+, Na^+ and K^+ for the exchange sites ($-SO_3^-$) causing the Na^+ and K^+ zones to move down the column. K^+ is more strongly retained than Na^+; thus the Na^+ zone moves down the column faster than the K^+ zone.

As originally conceived and carried out for many years, fractions of effluent were collected from the end of the column and analyzed for Na^+ and K^+. Then a plot was made of concentration vs. fraction number to construct a chromatogram. All this took a long time and made ion-exchange chromatography slow and awkward to use. However, it was soon realized that under a given set of conditions, all of the Na^+ would be in a single fraction of several milliliters and all of the K^+ could be recovered in a second fraction of a certain volume. Thus, under pre-determined conditions, each ion to be separated could be collected in a single fraction and then analyzed by spectroscopy, titration, etc., to determine the amount of each sample ion.

The situation regarding ion-exchange chromatography changed suddenly and drastically in 1975 when a landmark paper was published by Small, Stevens and Bauman [1]. Smaller and more efficient resin columns were used. But, more importantly, a system was introduced using a conductivity detector that made it possible to automatically detect and record the chromatogram of a separation. A new name was also introduced: ion chromatography. This name was originally applied to a patented system that used a conductivity detector in conjunction with a second ion-exchange column called a suppressor. This system will be described in detail a little later. However, the name "ion chromatography" is now applied to any modern, efficient separation that uses automatic detection.

In suppressed ion chromatography, anions are separated on a separator column that contains a low-capacity anion-exchange resin. A dilute solution of a base, such as sodium carbonate/sodium bicarbonate or sodium hydroxide is used as the eluent. Immediately following the anion-exchange "separator" column, a cation-exchange unit (called the suppressor) is used to convert the eluent to molecular carbonic acid, which has a very low conductivity. Also, the counterions of the sample anions are converted from sodium to hydrogen. The eluate from the suppressor unit then passes into a conductivity detector. If the sample ion pair is ionized to a reasonable extent, the sample anion (and the H^+ counterion) is detected by conductivity. An example of a state-of-the-art separation in the 1970s is shown in Fig. 1.1.

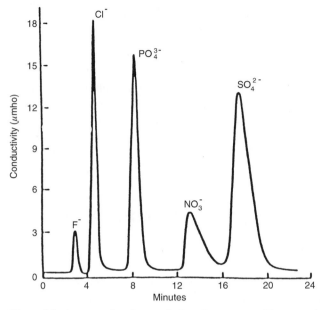

Figure 1.1. An example of an early ion chromatographic separation (From H. Small, *J. Chromatogr.*, 546, 3, 1991, with permission).

In the earlier instruments, the suppressor unit was a cation-exchange column of high capacity that had to be regenerated periodically. Newer suppressors contain ion-exchange membranes that can be regenerated continuously by flowing a solution of sulfuric acid over the outer membrane surface or by electrically generated acid.

Shortly after the invention of suppressed ion chromatography, Gjerde, Fritz and Schmuckler showed that ion chromatography separation and conductometric detection of anions and cations can be performed without the use of a suppressor unit [2–4]. Some early work was also performed by Harrison and Burge [5]. This technique was initially called single-column ion chromatography (SCIC) because only a single separation column is used, in contrast to the earlier suppressed systems in which two columns were used: a separator column and a suppressor column. However, non-suppressed ion chromatography now seems a more appropriate name.

For non-suppressed ion chromatography to be successful, the ion exchanger used in the separation column must have a low exchange capacity and a very dilute eluent must be used. In the separation of anions, the resin must have an exchange capacity between about 0.005 mequiv/g and 0.10 mequiv/g. Typical eluents are 1.0×10^{-4} M solutions of sodium or potassium salts of benzoic acid, hydroxybenzoic acid, or phthalic acid. These eluents are sufficiently dilute that the background conductivity is quite low. Most sample anions have a higher equivalent conductance than that of the eluent anion and can therefore be detected even when present in concentrations in the low parts per million range.

For the separation of cations, a cation exchange column of low capacity is used in conjunction with either a conductivity detector or another type of detector. With a conductivity detector, a dilute solution of nitric acid is typically used for separation of monovalent cations, and a solution of an ethylenediammonium salt is used for separation of divalent cations. Because both of these eluents are more highly conducting than the sample cations, the sample peaks are negative relative to the background (decreasing conductivity).

Shortly after the invention of suppressed ion chromatography, commercial instruments for its use were made available by the Dionex Corporation. Ion chromatography became an almost overnight sensation. It now became possible to separate mixtures such as F^-, Cl^-, Br^-, NO_3^- and SO_4^{2-} in minutes and at low ppm concentrations. Analytical problems that many never knew existed were described in an avalanche of publications.

1.3 Principles of Ion Chromatographic Separation and Detection

1.3.1 Requirements for Separation

The ion-exchange resins used in modern chromatography are smaller in size but have a lower capacity than older resins. Columns packed with these newer resins have more theoretical plates than older columns. For this reason, successful separations can now be obtained even when there are only small differences in retention times of the sample ions.

The major requirements of systems used in modern ion chromatography can be summarized as follows:
1. An efficient cation- or anion-exchange column with as many theoretical plates as possible.
2. An eluent that provides reasonable differences in retention times of sample ions.
3. A resin–eluent system that attains equilibrium quickly so that kinetic peak broadening is eliminated or minimized.
4. Elution conditions such that retention times are in a convenient range–not too short or too long.
5. An eluent and resin that are compatible with a suitable detector.

1.3.2 Experimental Setup

Anions in analytical samples are separated on a column packed with an anion exchange resin. Similarly, cations are separated on a column containing a cation-exchange resin. The principles for separating anions and cations are very similar. The separation of anions will be used here to illustrate the basic concepts.

A typical column used in ion chromatography might be 150×4.6 mm although columns as short as 50 mm in length or as long as 250 mm are also used. The column is carefully packed with a spherical anion-exchange resin of rather low exchange capacity and with a particle diameter of 5 or 10 μm. Most anion-exchange resins are functionalized with quaternary ammonium groups, which serve as the sites for the exchange of one anion for another.

The basic setup for IC is as follows. A pump is used to force the eluent through the system at a fixed rate, such as 1 mL/min. In the FILL mode a small sample loop (typically 10 to 100 μL) is filled with the analytical sample. At the same time, the eluent is pumped through the rest of the system, while by-passing the sample loop. In the INJECT mode a valve is turned so that the eluent sweeps the sample from the filled sample loop into the column. A detector cell of low dead volume is placed in the system just after the column. The detector is connected to a strip-chart recorder or a data-acquisition device so that a chromatogram of the separation (signal vs. time) can be plotted automatically. A conductivity- or UV-visible detector is most often used in ion chromatography. The hardware used in IC is described in more detail in Section 1.4.

The eluent used in anion chromatography contains an eluent anion, E^-. Usually Na^+ or H^+ will be the cation associated with E^-. The eluent anion must be compatible with the detection method used. For conductivity, the detection E^- should have either a significantly lower conductivity than the sample ions or be capable of being converted to a non-ionic form by a chemical suppression system. When spectrophotometric detection is employed, E^- will often be chosen for its ability to absorb strongly in the UV or visible spectral region. The concentration of E^- in the eluent will depend on the properties of the ion exchanger used and on the types of anions to be separated. Factors involved in the selection of a suitable eluent are discussed later.

1.3.3 Performing a Separation

To perform a separation, the eluent is first pumped through the system until equilibrium is reached, as evidenced by a stable baseline. The time needed for equilibrium to be reached may vary from a couple of minutes to an hour or longer, depending on the type of resin and eluent that is used. During this step the ion-exchange sites will be converted to the E^- form: $Resin-N^+R_3 \ E^-$. There may also be a second equilibrium in which some E^- is adsorbed on the resin surface but not at specific ion-exchange sites. In such cases the adsorption is likely to occur as an ion pair, such as E^-Na^+ or E^-H^+.

An analytical sample can be injected into the system as soon as a steady baseline has been obtained. A sample containing anions A_1^-, A_2^-, A_3^-,....,A_i^- undergoes ion-exchange with the exchange sites near the top of the chromatography column.

$$A_1^- \text{ (etc.)} + Res\text{-}E^- \ \rightleftarrows \ Res\text{-}A_1^- \text{ (etc.)} + E^-$$

injection peak can be eliminated by balancing the conductance of the injected sample with that of the eluent. Strasburg et al. studied injection peaks in some detail [6].

In suppressed anion chromatography, the effluent from the ion exchange column comes into contact with a cation-exchange device (Catex-H$^+$) just before the liquid stream passes into the detector. This causes the following reactions to occur.

Eluent: $Na^+OH^- + Catex\text{-}H^+ \rightarrow Catex\text{-}Na^+ + H_2O$

Chloride: $Na^+Cl^- + Catex\text{-}H^+ \rightarrow Catex\text{-}Na^+ + H^+Cl^-$

Bromide: $Na^+Br^- + Catex\text{-}H^+ \rightarrow Catex\text{-}Na^+ + H^+Br^-$

The background conductance of the eluent entering the detector is thus very low because virtually all ions have been removed by the suppressor unit. However, when a sample zone passes through the detector, the conductance is high due to the conductance of the chloride or bromide and the even higher conductance of the H$^+$ associated with the anion.

1.3.5 Detection

This effect can be used to practical advantage for the indirect detection of sample anions. For example, anions with little or no absorbance in the UV spectral region can still be detected spectrophotometrically by choosing a strongly absorbing eluent anion, E$^-$. An anion with a benzene ring (phthalate, *p*-hydroxybenzoate, etc.) would be a suitable choice. In this case, the baseline would be established at the high absorbance due to E$^-$. Peaks of non-absorbing sample anions would be in the negative direction owing to a lower concentration of E$^-$ within the sample anion zones.

Direct detection of anions is also possible, providing a detector is available that responds to some property of the sample ions. For example, anions that absorb in the UV spectral region can be detected spectrophotometrically. In this case, an eluent anion is selected that does not absorb (or absorbs very little).

1.3.6 Basis for Separation

The basis for separation in ion chromatography lies in differences in the exchange equilibrium between the various sample anions and the eluent ion. A more quantitative treatment of the effect of ion-exchange equilibrium on chromatographic separations is given later. Suppose the differences in the ion-exchange equilibrium are very small. This is the case for several of the transition metal cations (Fe^{2+}, Co^{2+}, Ni^{2+}, Cu^{2+}, Zn^{2+}, etc.) and for the trivalent lanthanides. Separation of the individual ions within these groups is very difficult when it is based only on the small differences in affinities of the ions for the resin sites. Much better results are obtained by using an eluent that complexes the sample ions to different extents. An equilibrium is set up between the sample cation, C^{2+}, and the complexing ligand, L^- in which species such as C^{2+}, CL^+, CL_2 and CL_3^- are formed. The rate of movement through the cation-exchange column is inversely proportional to α, the fraction of the element that is present as the free cation, C^{2+}.

1.4 Hardware

1.4.1 Components of an IC Instrument

This section describes the various components of an ion chromatography instrument, their function, and some general points for upkeep of the chromatograph. New IC users can use the information to understand how an instrument is built and to recognize the parts of the instrument that may need maintenance. The hardware is similar to that used for high pressure liquid chromatography (HPLC) but does have important differences. Readers who are familiar with HPLC will recognize the similarity and the differences to IC hardware [7–9].

Figure 1.3 shows a block diagram of the general components of an IC instrument. The hardware requirements for an IC include a supply of eluent(s), a high pressure pump (with pressure indicator) to deliver the eluent, an injector for introducing the sample into the eluent stream and onto the column, a column to separate the sample mixture into the individual components, an optional oven to contain the column, a detector to measure the analyte peaks as elute from the column and a data system for collecting and organizing the chromatograms and data.

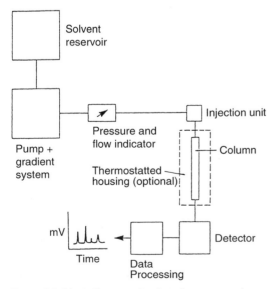

Figure 1.3. Block diagram of an ion chromatograph.

Everything on the high pressure side, from the pump outlet to the end of the column, must be strong enough to withstand the pressures involved. The wetted parts are usually made of PEEK and other types of plastics although other materials, such as sapphire, ruby, or even ceramics are used in the pump heads, check valves, and injector of the system. PEEK and other high performance plastics are the materials of choice for ion chromatography. Stainless steel can be used provided the system is properly conditioned to remove internal corrosion and the eluents that are used do not promote further corrosion. (Almost all IC eluents are not corrosive.) Stainless steel IC components are considered to be more reliable than those made from plastics, but require higher care. The stainless steel IC is normally delivered from the manufacturer pretreated so that corrosion is not present. The reader is advised to consult the IC instrument manufacturer for care and upkeep instructions.

1.4.2 Dead Volume

The dead volume of a system at the point where the sample is introduced (the injector) to the point where the peak is detected (the detection cell) must be kept to a minimum. Dead volume is any empty space or unoccupied volume. The presence of too much dead volume can lead to extreme losses in separation efficiency due to broadening of the peaks. Although all regions in the flow path are important, the most important region where peak broadening can happen is in the tubing and connections from the end of the column to the detector cell.

Of course there is a natural amount of dead volume in a system due to the internal volume of the connecting tubing, the interstitial spaces between the column packing beads and so on. But using small bore tubing (0.007 inch, 0.18 mm) in short lengths when making the injection to column and the column to detector connections is important. Also, it is important to make sure that the tubing end does not leave a space in the fitting when the connections are made. Dead volume from the pump to the injector should also be kept small to help to make possible rapid changes in the eluent composition in gradient elution.

Eluent entering the pump should not contain any dust or other particulate matter. Particulates can interfere with pumping action and damage the seal or valves. Material can also collect on the inlet frits or on the inlet of the column, causing pressure buildup. Eluents or the water and salt solutions used to prepare the eluents are normally filtered with a 0.2 or 0.45 µm nylon filter.

1.4.3 Degassing the Eluent

Degassing the eluent is important because air can get trapped in the check valve (discussed later in this section), causing the pump to lose its prime. Loss of prime results in erratic eluent flow or no flow at all. Sometimes only one pump head will lose its prime and the pressure will fluctuate in rhythm with the pump stroke. Another reason for removing dissolved air from the eluent is because air can result in changes in the effective concentration of the eluent. Carbon dioxide from air dissolved in water forms carbonic acid. Carbonic acid can change the effective concentration of a basic eluent, including solutions of sodium hydroxide, bicarbonate and carbonate. Usually, degassed water is used to prepare eluents and efforts should be made to keep exposure of eluent to air to a minimum after preparation.

Modern inline degassers are becoming quite popular. These are small devices that contain two to four channels. The eluent travels through these devices from the reservoirs to the pump. The tubing in the device is gas permeable and surrounded by vacuum. Helium sparging can also be used to degas eluents. Extended sparging may cause some retention shifts, so sparging should be reduced to a trickle after the initial few minutes of bubbling. Finally, it is best to change the eluents every couple of days to keep the concentration accurate.

1.4.4 Pumps

IC pumps are designed around an eccentric cam that is connected to a piston (Fig. 1.4). The rotation of the motor is transferred into the reciprocal movement of the piston. A pair of check valves controls the direction of flow through the pump head (discussed below). A pump seal surrounding the piston body keeps the eluent form leaking out of the pump head.

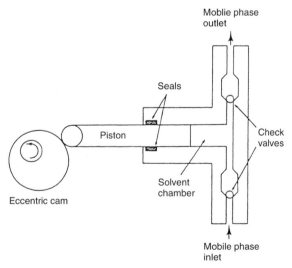

Figure 1.4. IC pump head, piston, and cam.

In single-headed reciprocating pumps, the eluent is delivered to the column for only half of the pumping cycle. A pulse dampener is used to soften the spike of pressure at the peak of the pumping cycle and to provide a eluent flow when the pump is refilling. Use of a dual head pump is better because heads are operated 180° out of phase with each other. One pump head pumps while the other is filling and vice versa.

The eluent flow rate is usually controlled by the pump motor speed although there are a few pumps that control flow rate by control of the piston stroke distance.

Figure 1.5 shows how the check valve works. On the intake stroke, the piston is withdrawn into the pump head, causing suction. The suction causes the outlet check valve to settle onto its seat while the inlet check valve rises from its seat, allowing eluent to fill the pump head. Then the piston travels back into the pump head on the delivery stroke. The pressure increase seals the inlet check valve and opens the outlet valve, forcing the eluent to flow out of the pump head to the injection valve and through the column. Failure of either of the check valves to sit properly will cause pump head failure and eluent will not be pumped. In most cases, this is due to air trapped in the valve so that the ball cannot sit properly. Flushing or purging the head usually takes care of this problem. Using degassed eluents is also helpful. In a few cases, particulate material can prevent sealing of the valve. In these cases, the valve must be cleaned or replaced. The pump manufacturer has instructions on how to perform this operation.

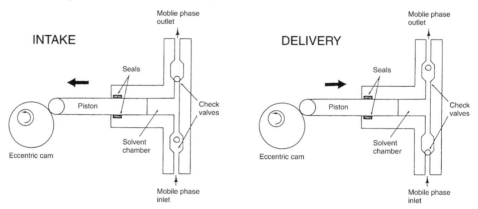

Figure 1.5. Check valve positions during intake and delivery strokes of the pump head piston.

1.4.5 Gradient Formation

Isocratic separations are performed with an eluent at a constant concentration of eluent buffer or salt solution. While it is desirable (simpler) to perform IC separations with single isocratic eluent, it is sometimes necessary to form a gradient of weak eluent to concentrated, strong eluent over a chromatographic run. This allows the separation of anions that may have a wide range of affinities for the column. Weakly adhering anions elute first and then, as the eluent concentration is increased, more strongly adhering anions can be eluted by the stronger eluent.

Figure 1.6 shows the two most popular methods for forming gradients. In the first method, flow from two high pressure pumps is directed into a high pressure mixing chamber. One pump contains a weak eluent while the other contains the stronger eluent. After the mixing chamber, the flow is directed to the injector and then on to the column. Controlling the relative pumping rate delivery of each pump forms the gradient. The total flow from the two pumps is constant. Starting with a high flow of the weak eluent pump and a low flow of the strong eluent pump, the gradient starts. Then, over the course of the chromatographic run, the relative flow rate of the strong eluent pump is increased while the flow rate of the weak eluent pump is decreased, keeping the total flow rate constant.

HIGH PRESSURE MIXING GRADIENT SYSTEM

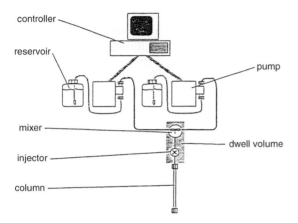

LOW PRESSURE MIXING GRADIENT SYSTEM

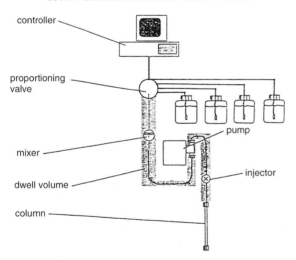

Figure 1.6. High-pressure mixing systems use two or more independent pumps to generate the gradient. Low-pressure mixing systems use a single pump with a proportioning valve to control composition. The advantages of high-pressure mixing are smaller dwell volumes and faster gradient formation. The advantages of low-pressure mixing are lower costs (single pump) and more versatile gradients (four solvents).

A more popular and less expensive method of forming gradients is by using a single pump and three or four micro-proportioning valves at the inlet of the pump. At low pressure, gradients can be formed from solvents A, B, C, and D (or any combination) by metering controlled amounts from the various eluent reservoirs into the pump. The composition in the low-pressure mixing chamber is controlled by timed proportioning valves. The time cycle remains constant throughout the gradient, but the time for any one of the solvent valves will vary. Only one eluent valve is open at any time. The gradient is formed by the relative time that the valves are open. At the start of the gradient, the valve connected to the weak eluent is open longest. As the gradient progresses, the valve connected to the strong eluent is open for longer and longer times, while the weaker eluent valve is open shorter times. The time cycle of the valve remains constant. Up to four valves are often available to give options for different types of gradients or the use of cleaning solutions. But generally, gradients are formed with just two of the valves.

1.4.6 Pressure

Column inlet pressures can vary from 500 psi up to perhaps a high of 3500 psi, with normal operating pressures around 1500 psi. The pressure limit on an IC is usually 4000 to 5000 psi, depending on the fittings and other hardware used. The eluent back-pressure is directly proportional to the eluent flow rate. Although still popular, psi (pounds per square inch) is gradually being replacing by more modern terms for pressure measurement, namely, 1 bar = 1 atm (atmospheres) = 14.5 psi = 10^5 Pa (Pascal).

1.4.7 Injector

The injection system may be manual or automated, but both rely on the injection valve. An injection valve is designed to introduce precise amounts of sample into the sample stream. The variation is usually less than 0.5 % precision from injection to injection. Figure 1.7 schematically represents the valve. It is a 6-port and 2-position device; one position is load and the other is inject. In the load position, the sample from the syringe or autosampler vial is pushed into the injection loop. The loop may be partially filled (partial loop injection) or completely filled (full loop injection) (Fig. 1.7). Partial loop injection depends on the precision filling of the loop with small known amounts of material. If partial loop injection is used, the loop must not be filled to more than 50 % of the total loop volume or the injection may not be precise. In full loop injection, the sample is pushed completely through the loop. Typical loop sizes are 10–200 µL. Normally at least a two-fold amount of sample is used to fill the loop with excess sample from the loop going to waste. At the same time that the sample loop is loaded with sample, the eluent travels in the by-pass channel of the injection valve and to the column. Injection of the sample is accomplished by turning the valve and placing the injection loop into the eluent stream. Usually the flow of the eluent is opposite to that of the loading sample into the loop. The injected sample travels to the head of the column as a slug of fluid. The ions in the sample interact with the column and the separation process is started with the eluent pushing the sample components down the column. Injection valves require periodic maintenance and usually have to be serviced after about 5000 injections. The manual for the instrument should be consulted for details on service.

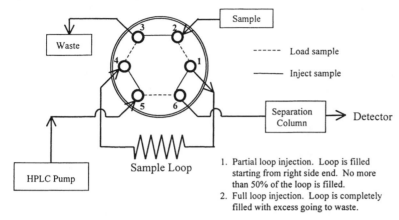

Figure 1.7. Schematic representation of partial- and full-loop injection methods.

1.4.8 Column Oven

The column oven is optional. Most IC separations are not dependent on the use of an oven (but see ion-exclusion and ligand-exchange chromatography for some exceptions). Nevertheless, an oven can be quite useful for high-sensitivity work. Conductivity is proportional to temperature. There is about a 2 % change in conductance per °C change in temperature. Conductivity detectors have temperature control, temperature compensation, or both. An oven can help to keep the temperature of the fluid, before the conductivity cell is reached, constant; this can help decrease the detector noise and decrease the detection limit of the instrument.

1.4.9 Column Hardware

IC columns are usually made of PEEK. Even the frits at the end of the column, which hold the column packing in place, are usually made of porous PEEK. The column lengths range from about 3 cm to 30 cm and the inside diameters range from about 1 mm to 7.8 mm i.d. Figure 1.8 shows the end of a column and the type of fitting used to connect the tubing to the column. Reusable PEEK fittings are used almost exclusively to connect tubing to columns and other instrument components. As stated earlier, the tubing should be bottomed out or pushed completely into the column end before the fitting is tightened, to ensure that there is no unnecessary dead volume in the connection.

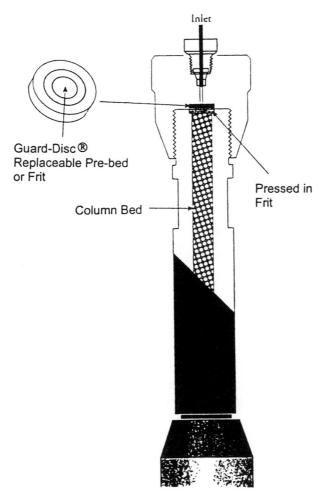

Figure 1.8. Representation of a typical IC column. The fittings, frits and body are normally PEEK. Some columns will contain a replaceable disc or frit at the top of the column to protect the column from particulates and contaminants (Courtesy of Transgenomic, Inc.).

1.4.10 Column Protection

Column protection not only extends the useful life of the separation column, but proper protection of the separation column can also result in more reliable analytical results over the lifetime of the column. Scavenger columns, located between the pump and injector, are one means of protecting the column. The scavenger removes particulate material that may be present in the eluent, but can also contain a resin to "polish" the eluent of any contaminant. An example is a chelating resin to remove metal contaminants. Besides protecting the separation column, scavenger columns may also improve detection of the analytes by reducing the background signal due to residual contaminants.

Most IC users do not use scavenger columns but rather prefer the use of guard columns, located directly in front of the separation column. Guard columns generally contain the same material as the separation column. Therefore, material that would be trapped and contaminate the separation column will instead get trapped by the guard column. Guard columns are changed when the separation of a standard is no longer acceptable and the column cannot be regenerated by the recommended procedures. Several guard columns may be used for protection over the lifetime of the separation column. Guard columns are generally smaller than the separation columns, but can add to the retention time of the separation. The Guard Disc column protection available on some IC columns is illustrated in Fig. 1.8. The disc is a packing material that is contained in a polymer matrix. The disc can contain up to 90 % packing material with the balance polymer fibrils to fix the resin. Particulate and dissolved contaminants are removed by the disc. The disc is, in essence, a thin, replaceable top section of the separation bed. Because the Guard Disc protection is thin, there is little impact on the separation.

1.4.11 Detection and Data System

The most common and useful detector for ion chromatography is conductivity; however, UV and other detectors can be quite useful. The types of detectors and their use are described in Chapter 4. The results of the chromatographic separation are generally displayed on a computer, although, in some older systems, recorders and integrators are used. The computer uses an A/D (analog to digital) board to convert the analog signal from the detector to digital data. The digital information is stored and manipulated to report results to the user.

The type of information that is most useful are the retention times of the various peaks and the peak areas (in a few cases, peak heights are used). Retention times are used to confirm the identity of the unknown peak by comparison with a standard. Peak area is compared to standards of known concentration to calculate analyte concentrations. This calculation can be performed by the use of a simple ratio:

$$\frac{\text{unknown concentration}}{\text{unknown peak area}} = \frac{\text{known concentration}}{\text{known peak area}}$$

therefore:

$$\text{unknown concentration} = \frac{\text{known concentration}}{\text{known peak area}} \times \text{unknown peak area}$$

It is usually better, however, to draw a calibration curve of known peak area vs. known concentration, to find the unknown peak area on the curve and to measure the unknown concentration on the axis.

For the data system to measure peak area, the baseline of the peak must be accurately drawn. The software program will attempt on its own to draw a baseline for the peak, but frequently the user must manually mark the baseline start and finish points to accurately draw the peak baseline.

1.4.12 Electrolytic Generation of Eluents

NaOH, KOH or other hydroxide eluents are desirable because of their low suppressed background conductivity. This leads to the ability to form eluent gradients with small shifts in the baseline. Also, low background conductivity can result in improved detection limits. For many workers, these advantages offset the disadvantages of limitations in control of selectivity and asymmetrical peaks for a few anions.

Pure hydroxide eluents are difficult to make, because of persistent contamination by carbon dioxide that is converted to carbonate. Carbonate is a much stronger eluting anion than hydroxide, and its presence can shift sample retention times to much shorter (and inconsistent) retention times. Carbonate will also cause baseline shifts when gradients are generated. In fact, a baseline shift during a hydroxide gradient is a good diagnostic indication that one or more of the eluent reservoirs contain carbon dioxide, bicarbonate, or carbonate anion.

The electrolytic generation of hydroxide eluent was first published by Dasgupta and coworkers [10, 11]. Rather than mixing reagents, the hydroxide eluent is formed electrochemically as it is being used and is introduced directly into the elution column from the generator. The system permits direct electrical control of the eluent concentration, and gradient chromatography is accomplished without mechanical proportioning. The system contains an anode and a cathode across which a DC current is passed. The reduction reaction at the cathode produces the hydroxide anion.

$$2\,H_2O + 2\,e^- \rightarrow 2\,OH^- + H_2 \uparrow \text{ (at cathode)}$$

A counterion to hydroxide is needed to conserve electric neutrality. Also, an oxidizing reaction occurs simultaneously at the anode. OH^- is electrolytically neutralized and O_2 is evolved.

$$H_2O - 2\,e^- \rightarrow 2\,H^+ + \tfrac{1}{2}\,O_2 \uparrow \text{ (at anode without NaOH)}$$

However, the feed solution for the anode also contains NaOH.

$$2\,Na^+ + 2\,OH^- + H_2O - 2\,e^- \rightarrow 2\,Na^+ + 2\,H_2O + \tfrac{1}{2}\,O_2 \uparrow \text{ (at anode with NaOH)}$$

The Na^+ and OH^- are combined to form the eluent through the use of an ion exchange membrane. A cation exchange membrane separating the anode from eluent flow allows the Na^+ to join the OH^- from the cathode. The membrane prevents passage of the OH^- that was originally associated with the Na^+ and therefore is available to combine with the H^+ formed at the anode, to produce H_2O.

A variation on the concept has been introduced by Dionex as the EG40 module [12]. In this case, KOH contained in a reservoir (labeled in Fig. 1.9 as K$^+$Electrolyte

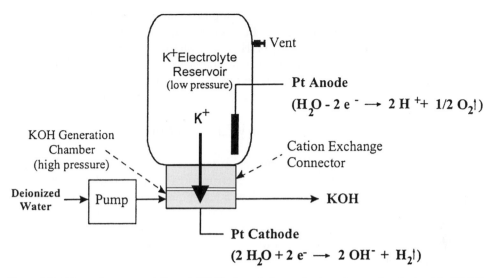

Figure 1.9. Schematic representation of EG40 electrolytic production of potassium hydroxide (KOH) eluent (courtesy of Dionex Corp).

reservoir (low pressure) is used rather than an NaOH feed solution. The process is the same, however, K^+ is generated, in effect, from the anode, because its counterion OH^- is consumed in the production of H^+. K^+ migrates across the cation exchange membrane to combine with OH^- formed at the cathode. Carbon dioxide is removed from the eluent stream en route to the EG40, to prevent contamination by carbonate. The electrolyte reservoir must be changed when the K^+Electrolyte is depleted.

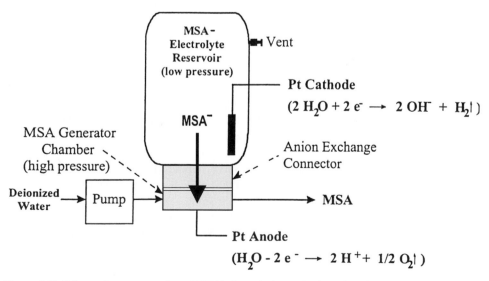

Figure 1.10. Schematic representation of EG40 electrolytic production of methanesulfonic acid (MSA) eluent (courtesy of Dionex Corp).

An analogous system can be used to generate methanesulfonic acid (MSA) eluent for the separation of cations (Fig. 1.10). In this case, the anode generates H^+ for eluent production. The cathode generates OH^- anion that combines with the H^+ in the MSA electrolyte reservoir. MSA^- anion migrates across the anion exchange membrane to combine with the H^+ eluent cation (maintaining electric neutrality).

1.5 Separation of Ions By Capillary Electrophoresis

It has been known for years that ions can be separated by differences in their rates of movement in an applied electric field (electrophoresis). Until fairly recently, electrophoresis was considered to be rather slow, a technique best reserved for separation of large organic ions and molecules. The realization that electrophoresis can be performed in a fused silica capillary has resulted in some dramatic changes. Now inorganic ions and small organic ions can be separated very quickly by this form of electrophoresis. Capillary zone electrophoresis (CZE), capillary electrophoresis (CE) and capillary ion electrophoresis (CIE) are the names most used to describe this type of separation.

In CE, ions are separated by their differences in mobility when a voltage of 15 to 30 kV is applied. Separations are very fast (usually <10 min) and the peaks are very sharp. Theoretical plate numbers of the order of 500 000 are not uncommon.

Separation of ions by capillary electrophoresis is covered in Chapter 10. The rationale for including CE in a book on ion chromatography is straightforward. If chromatography is defined broadly as separation resulting from differences in migration of the sample species, then capillary electrophoresis is indeed a type of chromatography. In CE, migration of the sample ions is the result of an applied electric field. In ion chromatography, migration of the sample ions is due to a flowing liquid phase.

1.6 Literature

A number of books on ion chromatography have been published. The first edition of Ion Chromatography (Fritz, Gjerde, and Pohlandt) was published in 1982 [13], followed by a second edition in 1987 [14]. The following authors have also written books on IC: Tarter, 1987 [15]; Small, 1989 [16]; Haddad and Jackson, 1990 [17]; Weiss, 1995 [18].

Many journal papers and short reviews have appeared on various aspects of ion chromatography. The Fundamental Reviews that appear every two years in Analytical Chemistry include new developments in ion chromatography under the heading "Liquid Chromatography: Theory and Practice" [19]. A collection of papers presented at the annual International Ion Chromatography Symposium has been published each year since 1989 [20, 21].

References

[1] H. Small, T. S. Stevens and W. S. Bauman, Novel ion-exchange chromatographic method using conductometric detection, *Anal. Chem.*, 47, 1801, 1975.

[2] D. T. Gjerde, J. S. Fritz and G. Schmuckler, Anion chromatography with low conductivity eluents, *J. Chromatogr.*, 186, 509, 1979.

[3] J. S. Fritz, D. T. Gjerde and R. M. Becker, Cation chromatography with a conductivity detector, eluents, II, *Anal. Chem.*, 52, 1519, 1980.

[4] D. T. Gjerde, G. Schmuckler and J. S. Fritz, Anion chromatography with low-conductivity eluents II, *J. Chromatogr.*, 187, 35, 1980.

[5] K. Harrison and D. Burge, Pittsburgh Conference paper 301, 1979.

[6] R. F. Strasburg, J. S. Fritz, J. Berkowitz and G. Schmuckler, Injection peaks in anion chromatography, *J. Chromatogr.*, 482, 343, 1989.

[7] S. Lindsay, High performance liquid chromatography, Wiley and Sons, New York, NY, 1987.

[8] R. W. Yost, L. S. Ettre and R. D. Conlon, Practical liquid chromatography, an introduction, Perkin Elmer, Norwalk, CT, 1980.

[9] E. L. Johnson and R. Stevenson, Basic liquid chromatography, Varian Associates, Walnut Creek, CA, 1978.

[10] D. L. Strong and P.K. Dasgupta, K. Friedman and J. R. Stillian, Electrodialytic eluent production and gradient generation in ion chromatography, *Anal. Chem.* 63, 480, 1991.

[11] D. L. Strong, C. U. Joung, K. K. Dasgupta, Electrodialytic eluent generation and suppression: ultralow background conductance suppressed anion chromatography, *J. Chromatogr.*, 546, 159, 1991.

[12] New products brochure EG40, Dionex Corp. Sunnyvale, CA, 1999.

[13] J. S. Fritz, D. T. Gjerde and C. Pohlandt, *Ion Chromatography*, Hüthig, Verlag, Heidelberg, 1982.

[14] D. T. Gjerde and J. S. Fritz, *Ion Chromatography*, Hüthig Verlag, Heidelberg, 2nd Ed., 1987.

[15] J. G. Tarter, *Ion Chromatography*, Dekker, New York, 1987.

[16] H. Small, *Ion Chromatography*, Plenum Press, New York, 1989.

[17] P. R. Haddad and P. E. Jackson, *Ion Chromatography, Principles and Applications*, Elsevier, Amsterdam, 1990.

[18] J. Weiss, Ion Chromatography, Second Ed., VCH, Weinheim, Germany, 1995.

[19] J. G. Dorsey, W. T. Cooper, B. A. Siles, J. P. Foley and H. G. Barth, Ion Chromatography, *Anal. Chem.*, 70, 613R, 1998.

[20] P. Jandik and R. M. Cassidy, Advances in Ion Chromatography, Century International, Franklin, MA. Vol. 1, 1989; Vol. 2, 1990.

[21] Various editors, Symposia on ion chromatography. *J. Chromatogr.*, Vols. 439, 482, 546, 602, 640, 671, 706, 739, 770; 1988–1997.

2 Historical Development of Ion-Exchange Separations

2.1 Introduction

The purpose of this chapter is to briefly trace some of the historical aspects of analytical separations involving ion-exchange resins. It is of at least passing interest to take a peek at ion-exchange chromatography as it was practiced "in the old days." But in the present time frame, we continue to write scientific history. There is always a certain logic in past developments that might provide ideas for future innovations.

Actually, chromatographic ion-exchange separations have been used for many years and an extensive literature exists. Although older ion-exchange separations are often slow and cumbersome by modern standards, these procedures illustrate a number of clever and useful approaches. There is no reason why many of these classical separations cannot be adapted to more modern chromatography technology.

Perhaps the major dividing line between the old and new in ion-exchange separation is the type of detection used. In classical procedures, numerous fractions of effluent from the ion-exchange column were collected. The amount of sample ions in each fraction was determined by titration, spectrometry, or another form of analysis. By plotting the concentration of sample ion as a function of fraction number, a rudimentary chromatogram could be constructed. The time required for such an operation certainly limited the enthusiasm an analyst might have for this type of separation. However, once the chromatogram for a given type of separation had been established, it became possible to collect only a single, larger fraction of effluent for each item to be determined.

The ability to collect a single fraction that contains all of the separated sample ion permits the use of step gradients. In this mode, conditions are adjusted so that an "all-or-nothing" situation prevails. A sample ion either sticks onto the ion exchange column or it passes quickly through. Conditions are selected so that only one ion type will pass through the column while the other sample ions are strongly retained and form a tight band at the top of the column. Then the eluent is changed so that a second ion is rapidly eluted, while the others remain tightly stuck. Frequently, several gradient steps can be performed, to elute different sample ions at each step.

Step gradient elution is also well suited for group, or ionic class separations. For example, a chelating resin might be specific for the platinum-group metals. With this resin, a step-gradient scheme could be devised to take up, and later elute, only the selected group of metal ions.

Step elutions are often easy to achieve. For example, if metal cation A is converted to a neutral or anionic complex, it will pass quickly through a cation exchange column. A second sample cation B that is not complexed will be strongly taken up by the column. When all of A has been washed off the column, B can then be eluted by a step change to an appropriate eluent. The detection requirements are the same as for other gradients. A steady baseline must be achieved; this usually requires a selective detection method.

Step gradient methods can be used to concentrate trace amounts of uncomplexed sample ions on a very small column. Often up to several liters of sample solution can be concentrated.

Automatic detection of separated metal ions began at least as early as 1971 when iron (III) was eluted by hydrochloric acid from a column and detected by a UV detection cell [1]. Seymour and Fritz described a Z-type flow-through cell for UV detection at 270 nm in 1973 [2].

Modern ion chromatography was in a sense "born" in 1975 when Small, Stevens and Bauman [3] devised a new system that made it possible to use a conductivity detector. In this paper, 0.01 to 0.02 M hydrochloric acid was used as the eluent in the separation of alkali metal cations. A "stripper" column (later called a suppressor column) below the separator column, containing an anion-exchange resin in the -OH$^-$ form, was used to convert the H$^+$Cl$^-$ to water.

The alkali-metal sample ions were converted to the highly conducting metal hydroxide, e.g.,

$$Na^+Cl^- + Anex\text{-}OH^- \rightarrow Anex\text{-}Cl^- + Na^+OH^-$$

A solution of sodium phenolate, and, later, a mixture of sodium carbonate and bicarbonate, were used as the eluent for the separation of anions on an anion-exchange column. A cation-exchange stripper column was used to reduce the background conductance of the eluent and to enhance the conductance of sample anions such as chloride.

$$Na^+OC_6H_5^- + Catex\text{--}H^+ \rightarrow Catex\text{--}Na^+ + HOC_6H_5$$

$$Na^+Cl^- + Catex\text{--}H^+ \rightarrow Catex\text{--}Na^+ + H^+Cl^-$$

The Dionex Co. was formed to commercialize this invention, and ion chromatography, as the new technique was named, became an overnight sensation.

In 1979, Gjerde and Fritz [4] published a study in which anion-exchangers with exchange capacities ranging from 0.04 to 1.46 mequiv/g were prepared from macroporous polystyrene-DVB resins. The capacities were controlled by varying the time of chloromethylation and then alkylating the chloromethyl groups.

$$Resin-OCH_2Cl + R_3N \rightarrow Resin-OCH_2N^+R_3$$

The distribution ratios of inorganic anions were found to decrease with decreasing resin exchange capacity. Simply by using an anion exchange resin of low exchange capacity, it was now possible to use a much lower ionic concentration in the eluent for anion exchange chromatography.

Gjerde, Fritz and Schmuckler [5] devised a form on anion chromatography in 1979 that used a conductivity detector but did not require the use of a second stripper, or suppressor column. The anion-exchange resin used had a capacity of only 0.007–0.07 mequiv/g and only a very dilute solution (ca. 10^{-4} M) of an organic acid salt was needed in the eluent. The organic anion (benzoate or phthalate) had a much lower equivalent conductance than the typical inorganic anions to be separated. Thus, the detection was based on an increase in conductance when a sample anion passed through the detector. While the detection sensitivity was not as good as the suppressed system, it was quite adequate for most separations. The nonsuppressed system (called single-column ion chromatography at the time) also allowed greater flexibility in the eluent ions that could be used.

The nonsuppressed method for anion chromatography was followed quickly by a similar method for cations [6]. A mixture of Li^+, Na^+, NH_4^+, K^+, Rb^+, and Cs^+ was separated in less than 10 min with a blend of 0.017 mequiv/g and unfunctionalized cat-

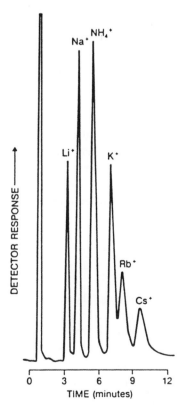

Figure 2.1. Separation of alkali-metal cations and ammonium cation on a low-capacity cation exchange column with a conductivity detector. Eluent: 1.0 mM nitric acid. Column: 350 × 2.0 mm packed with 0.059 mequiv/g cation-exchange resin.

ion exchange resins, an eluent containing 1.25×10^{-3} M nitric acid and a conductivity detector (Fig. 2.1). The sample peaks were in the direction of decreasing conductivity due to the partial replacement of the highly conductive H^+ by the sample cations of lower conductivity.

Development of the ion-chromatographic methods that used a conductivity detector was accompanied by a significant increase in chromatographic efficiency. The ion-exchange materials were of much smaller and more uniform size and column packing efficiency was also improved. The changes that occurred were not unlike those in partition chromatography when it went from "liquid chromatography" to "high-performance liquid chromatography" (HPLC).

2.2 Separation of Cations

2.2.1 Cation Separations Based On Affinity Differences

Strelow and his coworkers have published extensive data relating to the selectivity of a sulfonated polystyrene cation exchanger for various cations in acidic solution [7]. The equilibria of cations in hydrochloric, nitric or sulfuric acid solutions with a cation exchanger involves complexation in some cases as well as competition between H^+ and the metal cation for the exchange sites. For example, mercury(II) and cadmium(II) form chloride complexes even in dilute solutions of hydrochloric acid. Selectivity data in perchloric acid probably give the best indication of true ion-exchange selectivity, because the perchlorate anion has almost no complexing properties with metal cations.

In general, cations with a 3+ charge are more strongly retained by a cation exchanger than cations with a 2+ charge, and ions with a 2+ charge are retained more strongly than those with a 1+ charge. Fritz and Karraker [8] were able to separate metal cations into groups according to their charge. Most divalent metal cations were eluted with a 0.1 M solution of ethylenediammonium perchlorate. Then the trivalent metal ions remaining on the column were eluted with 0.5 M ethylenediammonium perchlorate. Bismuth(III) and zirconium(IV) remained quantitatively on the cation exchange column. The use of the 2+ ethylenediammonium ion permitted a lower concentration to be used than would have been the case with a H^+ eluent.

2.2.2 Cation Separations with Complexing Eluents

Several inorganic acids exhibit a complexing effect for metal ions. The complexing acids include HF, HCl, HBr, HI, HSCN, and H_2SO_4. The complexed metal ions are converted into neutral or anionic complexes and are rapidly eluted, while the other cations remain on the cation-exchange column.

The data for hydrochloric acid [9] indicate selective complexing between metal cations and the chloride ion. For example, cadmium(II) has a distribution coefficient of 6.5 in 0.5 M hydrochloric acid, but a $D = 101$ in 0.5 M perchloric acid.

Calcium(II), which shows no appreciable complexing, has a distribution coefficient of 147 in 0.5 M perchloric acid and 191 in 0.5 M hydrochloric acid. Strelow, Rethemeyer, and Bothma [10] also reported data for nitric and sulfuric acids that showed complexation in some cases. Mercury(II), bismuth(III), cadmium(II), zinc(II), and lead(II) form bromide complexes and are eluted in the order given in 0.1 to 0.6 M hydrobromic acid [11]. Most other metal cations remain on the column. Aluminum(III), molybdenum(VI), niobium(V), tin(IV), tantalum(V), uranium(VI), tungsten(VI), and zirconium(IV) form anionic fluoride complexes and are quickly eluted from a hydrogen-form cation-exchange column with 0.1 to 0.2 M HF [12].

An eluent containing only 1 % hydrogen peroxide in dilute aqueous solution will form stable anionic complexes with several metal ions. Fritz and Abbink [13] were able to separate vanadium(IV) or (V) from 25 metal cations including the separation of vanadium (V) from 100 times as much iron(III).

Strelow [14] used hydrogen peroxide and sulfuric acid to separate titanium(IV) from more than 20 cations by cation exchange. Fritz and Dahmer [15] separated molybdenum(VI), tungsten(VI), niobium(V) and tantalum(V) as a group from other metals by adding dilute hydrogen peroxide to the sample solution and passing it through a cation-exchange column.

Most of the eluents listed above are volatile upon heating and do not interfere with colorimetric, titrimetric or other methods for chemical determination of the metal ions separated. For the most part, group separations, rather than separation of individual metal ions, are obtained and only a short ion-exchange column is needed. Another valuable "all or nothing" group separation uses an eluent consisting of 0.1 M tartaric acid and 0.01 M nitric acid [16]. Antimony(V), molybdenum(VI), tantalum(V), tin(IV), and tungsten(VI) form tartrate complexes in this acidic medium but lead(II) and many other metal cations are not complexed and are retained by the cation exchanger. Samples containing tin(IV) must be added to the column in the tartrate solution.

In a few cases an eluent containing an organic complexing reagent has been used successfully for the chromatographic separation of several metal ions. A notable example is the separation of individual rare earth ions with a solution of 2-hydroxyl-isobutyric acid as the eluent [17]. However, such separations necessitate careful equilibration of the column to maintain a desired pH. Sometimes gradient elution is used, and either the pH or the eluent concentration is changed.

2.2.3 Effect of Organic Solvents

Metal cations usually form complexes with inorganic anions much more readily in organic solvents than in water. For example, the pink cobalt(II) cation requires around 4 or 5 M aqueous hydrochloric acid to be converted to a blue cobalt(II) chloride anion. In a predominantly acetone solution, the intensely blue cobalt(II) is formed in very dilute hydrochloric acid. Thus, the scope of ion-exchange group separations is increased greatly by carrying out separations in a mixture of water and an organic solvent.

Fritz and Rettig [18] showed that zinc(II), iron(III), cobalt(II), copper(II) and manganese(II) can be separated from each other on a short cation-exchange column with eluents containing a fixed, low concentration of HCl and increasing the water-acetone proportion from 40 % to 95 % acetone in steps. Later Strelow et al. [19] published extensive lists of metal-ion distribution coefficients in water/acetone/hydrochloric acid systems.

Korkisch and co-workers have studied the effect of ethanol, acetic acid, ethylene glycol, and many other solvents upon the ion-exchange behavior of metal ions in systems containing hydrochloric and other complexing acids [20].

The selectivity of low capacity cation columns for monovalent ions can be adjusted by the addition of an organic modifier to the eluent. Using a nitric acid eluent of pH 2.5, for example, the elution order for monovalents is Li^+, Na^+, NH_4^+, K^+. Simple amines elute in the order of the carbon number, after NH_4^+, with the result that $(CH_3)NH_3^+$ (methylammonium) can co-elute with potassium. In most cases, this co-elution is of little significance, because potassium and methylammonium are not often in the same sample. However, where the analysis of either of these species in the presence of the other is desired, the selectivity can be modified by the addition of 40 % methanol to the eluent [21]. The methanol causes the potassium to elute later but does not affect the elution time of methylammonium, and potassium peaks are now resolved from each other.

2.3 Separation of Anions

2.3.1 Separation of Anions with the Use of Affinity Differences

There are considerable differences in the selectivity coefficients of common inorganic anions for anion-exchange resins as shown by the data in Section 3.5. Classical separations based on these affinity differences have been achieved, although they were often quite slow. However, excellent separations of anion mixtures can be obtained by modern anion chromatography.

2.3.2 Anion Separations Involving Complex Formation

Kraus and Nelson, working at Oak Ridge National Laboratory in the USA, found that in aqueous hydrochloric acid solutions a number of metal ions form anionic complexes and are strongly taken up by anion-exchange resins. For most metal ions, a plot of the D value of several thousand is attained. An illustration of such plots for most of the metallic elements in the periodic table was published by Kraus and Nelson in 1956 [22], and has been reproduced in Fig. 2.2.

Separations are generally achieved by adding the sample to an anion-exchange column in rather concentrated hydrochloric acid and eluting the nonsorbed metal ions with eluent of the same HCl concentration. Then the sorbed metal ions are eluted one at a time by stepwise reduction of the HCl strength of the eluent. Figure 2.3 illustrates one of the many practical separations published by Kraus and his co-workers [23].

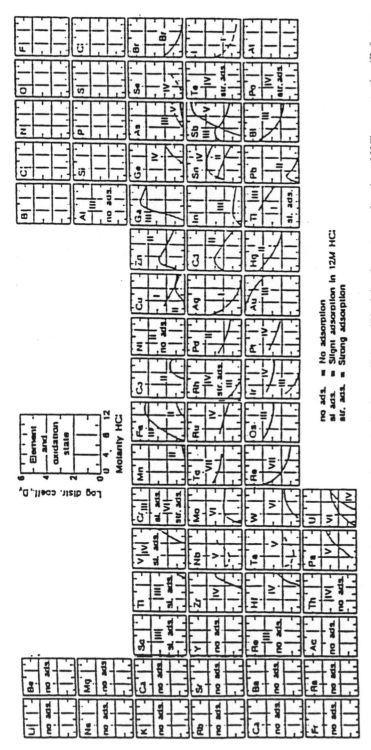

Figure 2.2. Classical ion-exchange separations. Anion-exchange distribution coefficients (*D*) of the elements as a function of HCl concentration (Reference [22], with permission).

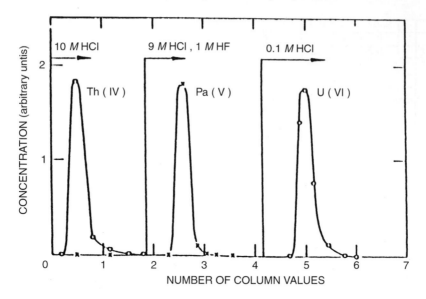

Figure 2.3. Separation of metal ions on 200-mesh Dowex 1×10 anion exchange resin (Reference [24], with permission).

In a similar manner, elements that form anionic fluoride complexes can be separated from others and from each other on an anion exchanger by eluting with eluents containing HF plus HCl [24, 25]. Extensive studies of metal ion behavior on anion-exchange columns have also been carried out with eluents containing mixed H_2SO_4/HF [26, 27].

Anion-exchange distribution coefficients for most metallic elements in sulfuric acid solution have been measured [28, 29]. Uranium(VI), thorium(IV), molybdenum(VI), and a few other elements are retained selectively from such solutions. Thorium(IV) is taken up selectively by anion-exchangers from approximately 6 M nitric acid [30].

2.3.3. Effect of Organic Solvents

As noted earlier for cation-exchange separations, operating in a predominately organic solvent greatly improves the ability of metal ions to form complexes with halide and pseudo-halide anions. Such complexes generally are taken up strongly by an anion-exchange resin. Korkisch, Fritz, Strelow, and others have published extensively on anion-exchange separations in partly nonaqueous solutions. Korkisch and Hazan [31] described a method to separate metal ions that form chloride complexes from those that do not. The method uses an eluent consisting of 90 to 95 % methanol/0.6 M hydrochloric acid and requires only a short anion-exchange column. The metal ions studied are either retained as a sharp band or quickly pass through the column. Thus, we have an "all or nothing" situation and excellent group separations are obtained.

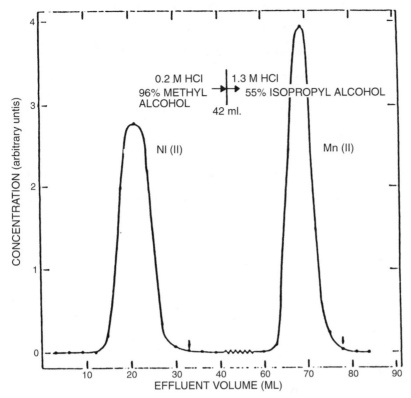

Figure 2.4. Separation of nickel(II) and manganese(II) on a 6.0 × 2.2 cm column containing Dowex 1 × 8 resin, with partly nonaqueous eluents (Reference [31], with permission).

Chromatographic separations of individual ions are also possible and many have been published. An example is shown in Fig. 2.4. Ion exchange in nonaqueous and mixed media has been reviewed [32].

Systems containing dimethylsulfoxide, methanol and hydrochloric acid have been studied for the anion-exchange behavior of 26 elements [33]. Numerous separations of two- to four-component mixtures of metal ions were carried out with quantitative results.

References

[1] M. D. Seymour, J. P. Sickafoose and J. S. Fritz, Application of forced-flow liquid chromatography to the determination of iron, *Anal. Chem.*, 43, 1734, 1971.
[2] M. D. Seymour and J. S. Fritz, Rapid, selective method for lead by forced-flow liquid chromatography, *Anal. Chem.*, 45, 1632, 1973.
[3] H. Small, T. S. Stevens and W. G. Bauman, Novel ion exchange chromatographic method using conductimetric detection, *Anal. Chem.*, 47, 1801, 1975.
[4] D. T. Gjerde and J. S. Fritz, Effect of capacity on the behavior of anion-exchange resins, *J. Chromatogr.*, 176 199, 1979.
[5] D. T. Gjerde, J. S. Fritz and G. Schmuckler, Anion chromatography with low conductivity eluents, I, *J. Chromatogr.*, 186 509, 1979; II, *J. Chromatogr.*, 187, 35, 1980.

[6]   J. S. Fritz, D. T. Gjerde and R. M. Becker, Cation chromatography with a conductivity detector, *Anal. Chem.*, 52, 1519, 1980.

[7]   F., W. E. Strelow and H. Sondorp, Distribution coefficients and cation-exchange selectivities of elements with AG50W-X8 resins in perchloric acid, *Talanta*, 19, 1113, 1972.

[8]   J. S. Fritz and S. K. Karraker, Ion exchange separation of metal cations, *Anal. Chem.*, 32, 957, 1960.

[9]   F. W. E. Strelow, An ion exchange selectivity scale of cations based on equilibrium distribution coefficients, *Anal. Chem.*, 32, 1185, 1960.

[10]   F. W. E. Strelow, R. Rethemeyer, and C. J. C. Bothma, Ion exchange selectivity scales for cations in nitric acid and sulfuric acid with a sulfonated polystyrene resin, *Anal. Chem.*, 37, 106, 196.

[11]   J. S. Fritz and B. B. Garralda, Cation exchange separation of metal ions with hydrobromic acid, *Anal. Chem.* 34, 102, 1962.

[12]   J. S. Fritz, B. B. Garralda and S. K. Karraker, Cation exchange separation of metal ions by elution with hydrofluoric acid, *Anal. Chem.* 33, 882, 1961.

[13]   J. S. Fritz and J. E. Abbink, Cation exchange separation of vanadium from metal ions, *Anal. Chem.*, 34, 1080, 1962.

[14]   F. W. E. Strelow, Separation of titanium from rare earths, beryllium, niobium, iron, aluminum, thorium, magnesium, manganese and other elements by cation exchange chromatography, *Anal. Chem.*, 35, 1279, 1963.

[15]   J. S. Fritz and L. H. Dahmer, Cation exchange separation of molybdenum, tungsten, niobium and tantalum from other metals, *Anal. Chem.*, 37, 1272, 1965.

[16]   F. W. E. Strelow and T. N. van der Walt, Separation of lead from tin, antimony, niobium, tantalum, molybdenum and tungsten by cation exchange chromatography in tartaric-nitric acid mixtures, *Anal. Chem.* 47, 2272, 1975.

[17]   J. N. Story and J. S. Fritz, Forced-flow chromatography of the lanthanides employing continuous in-stream detection, *Talanta* 21, 894, 1974.

[18]   J. S. Fritz and T. A. Rettig, Cation exchange in acetone-water-hydrochloric acid, *Anal. Chem.* 34, 1562, 1962.

[19]   F. W. E. Strelow, A. H. Victor, C. R. van Zyl, and C. Eloff, Distribution coefficients and cation exchange behavior of elements in hydrochloric acid-acetone, *Anal. Chem.*, 43, 870, 1971.

[20]   J. Korkisch, Modern methods for the separation of rarer metal ions, Pergamon, Oxford 1969.

[21]   Wescan Instruments, Inc., Santa Clara, CA, "Wescan Ion Analyzer #6" (1983).

[22]   K. A. Kraus and F. Nelson, Proc. First U. N. Int. Conf. on Peaceful Uses of Atomic Energy, 7, 113, 1956.

[23]   K. A. Kraus, G. E. Moore and F. Nelson, Anion exchange studies. XXI. Th(IV) and U(IV) in hydrochloric acid. Separation of thorium, protoactinium and uranium, *J. Am. Chem. Soc.*, 78, 2692, 1956.

[24]   F. Nelson, R. M. Rush and K. A. Kraus, Anion exchange studies. XXVII. Adsorbability of a number of elements in HCl-HF solutions, *J. Am. Chem. Soc.*, 82, 339, 1960.

[25]   J. B. Headridge and E. J. Dixon, The analysis of complex alloys with particular reference to niobium, tantalum and tungsten, *Analyst*, 87, 32, 1962.

[26]   E. A. Huff, Anion exchange study of a number of elements in nitric-hydrofluoric acid mixtures, *Anal. Chem.*, 36, 1921, 1964.

[27]   L. Danielsson, Adsorption of a number of elements from HNO_3-HF and H_2SO_4-HF solutions by cation and anion exchange, *Acta Chem. Scand.*, 19, 1859, 1965.

[28]   L. Danielsson, Adsorption of a number of elements from sulfuric acid solutions by anion exchange, *Acta Chem. Scand.*, 19, 670, 1965.

[29]   F. W. E. Strelow and C. J. C. Bothma, Anion exchange and a selectivity scale for elements in sulfuric acid media with a strongly basic resin, *Anal. Chem.*, 39, 595, 1967.

[30]   J. S. Fritz and B. B. Garralda, Anion exchange separation of thorium using nitric acid, *Anal. Chem.*, 34, 1387, 1962.

[31]   J. Korkisch and I. Hazan, Anion exchange behavior of uranium, thorium, the rare earths and various other elements in hydrochloric acid-organic solvent media, *Talanta*, 11, 1157, 1964.

[32]   W. R. Heumann, Ion exchange in nonaqueous and mixed media, Critical Revs. in *Anal. Chem.*, 2, 425, 1971.

[33]   J. S. Fritz and Marcia Lehoczky Gillette, Anion-exchange separation of metal ions in dimethyl-sulfoxide-methanol-hydrochloric acid, *Talanta*, 15, 287, 1968.

3 Ion-Exchange Resins

3.1 Introduction

A cation exchanger is a solid particulate material with negatively charged functional groups arranged to interact with ions in the surrounding liquid phase. For convenience, we will often refer to a cation exchanger as a "catex". The most common type of catex contains sulfonic acid groups. Cross-linked polystyrene particles are converted to a catex by sulfonation with concentrated sulfuric acid.

$$\text{Res}-\langle\bigcirc\rangle + H_2SO_4 \rightarrow \text{Res}-\langle\bigcirc\rangle-SO_3^-H^+ + H_2O \qquad (3.1)$$

In this equation, Res denotes resin or polymer. Silica-based cation exchangers are generally prepared by reacting silica particles with an appropriate chlorosilane or methoxysilane. A common type of silica catex has the structure:

$$\text{Silica} - O - \overset{|}{\underset{|}{Si}} - CH_2-\langle\bigcirc\rangle - SO_3^-H^+$$

In both of these materials, the sulfonate group is chemically bonded to the solid matrix. However, the H^+ is attracted electrostatically to the $-SO_3^-$ group and can undergo exchange reactions with other ions in solution. For example:

$$\text{Solid--SO}_3^-H^+ + Na^+ \;\rightleftarrows\; \text{Solid--SO}_3^-Na^+ + H^+ \qquad (3.2)$$

$$2\,\text{Solid--SO}_3^-Na^+ + Mg^{2+} \;\rightleftarrows\; (\text{Solid--SO}_3^-)_2\,Mg^{2+} + 2H^+ \qquad (3.3)$$

The physical form of the catex is such that ions from the surrounding solution can readily traverse through the solid to come into contact with the interior as well as the surface sulfonate groups.

The exchange reactions (Eqs. 3.2 and 3.3) are reversible and are subject to the laws of chemical equilibrium. Most monovalent metal ions are more strongly held by the catex than H^+. Cations of a higher charge are usually retained more strongly than those of lower charge. Ion-exchange equilibria are treated in more detail in Chapter 5.

Anion-exchangers (sometimes abbreviated as "anex") may be polymer-based (commonly polystyrene or polyacrylate) or silica-based. Although many types of anion exchangers are available for ion chromatography, the "strong-base" type with quaternary ammonium functional groups is the most common. These normally come in the chloride form, for example, Solid–$N^+R_3Cl^-$.

The alkyl R groups are usually methyl or a methyl with one or two hydroxyethyl groups, $-CH_2CH_2OH$. The positively charged quaternary ammonium groups are chemically bonded to the solid particles while the chloride groups are able to undergo ion exchange with other anions.

3.2 Polymeric Resins

3.2.1 Substrate and Cross-Linking

A variety of polymeric substrates can be used in ion-exchange synthesis, including polymers of esters, amides, and alkyl halides. But resins based on styrene–divinylbenzene copolymers are probably the most widely used ion exchangers. The polymer is schematically represented in Fig. 3.1. The resin is made up primarily of polystyrene; however, a small amount of divinylbenzene is added during the polymerization to "cross-link" the resin. This cross-linking confers mechanical stability upon the polymer bead and also dramatically decreases the solubility of the polymer by increasing the molecular weight of the average polymer chain length. Typically, 2 to 25 % weight of the cross-linking compound is used for microporous resins and up to 55 % weight cross-linking for macroporous resins. In many cases, the resin name will indicate the cross-linking of the material. For example, a Dowex 50 × 4 cation exchanger contains 4 % divinylbenzene polymer.

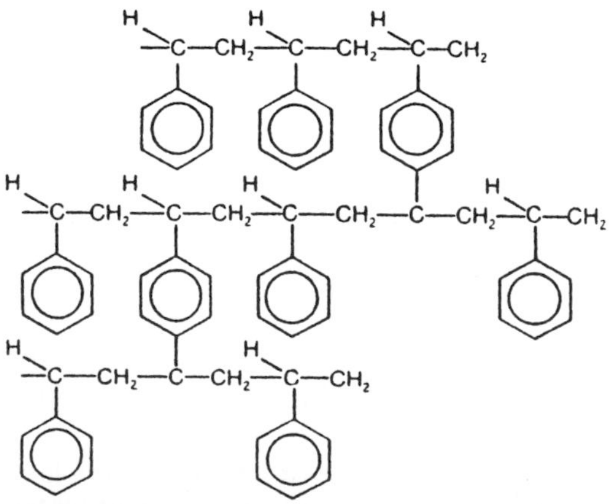

Figure 3.1. Schematic representation of a styrene–divinylbenzene copolymer. The divinylbenzene "cross-links" the linear chain of the styrene polymer. A high percentage of divinylbenzene produces a more rigid polymer bead.

3.2.2 Microporous Resins

The starting material for cation- and anion-exchange "polymer" resins can be classified either as microporous or macroporous. Most classical work has been done with microporous ion-exchange resins. Microporous substrates are produced by a suspension polymerization in which styrene and divinylbenzene are suspended in water as droplets. The monomers are kept in suspension in the reaction vessel through rapid, uniform stirring. Addition of a catalyst such as benzoyl peroxide initiates the polymerization. The resulting beads are uniform and solid but are said to be microporous. The size distribution of the beads is dependent on the stirring rate, that is, faster stirring produces smaller beads. The beads swell but do not dissolve when placed in common hydrocarbon solvents. After the resin is functionalized (the ion-exchange functional groups are attached to the polymer), the bead is considerably more polar. Depending on the relative number of functional groups, polar solvents such as water will now swell the ion-exchange resin. However, nonpolar solvents will tend to dehydrate the bead and cause it to shrink.

The extent of ion-exchange resin hydration will also depend on the ionic form of the resin. Ion-exchange resin beads with very little cross-linking are soft and tend to swell or shrink excessively when converted from one ionic form to another. However, the amount of cross-linking used in resin synthesis is still based on a compromise of resin performance. Microporous polystyrene resins usually contain about 8 % divinylbenzene. Gel-type resins with a high cross-linking tend to exclude larger ions, and the diffusion of ions of ordinary dimensions within the gel may be slower than might be desired. Resins with cross-linking lower than about 2 % are too soft for most column work.

3.2.3 Macroporous Resins

Macroporous resins (sometimes called macroreticular resins) are prepared by a special suspension polymerization process. Again, as with microporous resins, the polymerization is performed while the monomers are kept as a suspension of a polar solvent. However, the suspended monomer droplets also contain an inert diluent that is a good solvent for the monomers, but not for the material that is already polymerized. Thus, resin beads are formed that contain pools of diluent distributed throughout the bead matrix. After polymerization is complete, the diluent is washed out of the beads to form the macroporous structure. The result is rigid, spherical resin beads that have a high surface area.

Rather than using an inert solvent to precipitate the copolymer and form the pores, the polymerization may be carried out in the presence of an inert solid agent such as finely divided calcium carbonate to create the voids within the bead. Later, the solid is also extracted from the copolymer. Both of these polymerization processes create large (although probably different) inner pores. The average pore diameter can be varied within the range of 20 Å to 500 Å.

The final resin bead structure of a macroreticular resin contains many hard microspheres interspersed with pores and channels. Because each resin bead is really made up of thousands of smaller beads (something like a popcorn ball), the surface area of macroporous resins is much higher than that of microporous resins. A gel resin has a (calculated) surface area of less than 1 m^2/g. However, macroporous resin surface areas range from 25 to as much as 800 m^2/g.

Macroporous resins are remarkably rigid because of the large amounts of cross-linking agents normally used in the synthesis. Such resins are particularly advantageous for performing ion-exchange chromatography in organic solvents since changing solvent polarity does not swell or shrink the resin bed as it might for a gel-type resin. But the high cross-linking does not inhibit the ion-exchange process as it does in gel resins because the resins have pores and channels that are easily penetrated by the ions.

3.2.4 Chemical Functionalization

Ion exchangers are created by chemically introducing suitable functional groups into the polymeric matrix. In a few instances, monomers are functionalized first, and then they are polymerized into beads. An attractive feature of the aromatic copolymer used in many ion exchangers is that it can be modified easily by a wide variety of chemical reactions. More recently, some ion-exchange substrates have been polymers of esters (polymethacrylate) or amides. The reaction solvent is important in ion-exchange synthesis. In many cases, gel substrates must first be swollen in the reaction solvent to achieve complete functionalization of the resin (We shall see that complete functionalization is not desired in many resins in ion chromatography). The reaction solvent does not appear to be as critical in ion-exchange synthesis of macroporous substrates. Their surface is already "exposed" and ready to be converted to an ion exchanger.

Weak-acid cation-exchange resins that contain a carboxylic acid ion-exchange group are sometimes used. However, the most popular type of cation exchanger is made by introducing a sulfonic acid functional group. Resins with the sulfonic acid group are said to be strong acid ion exchangers. Sulfonation reactions are performed by treating the polystyrene resin with concentrated sulfuric acid. Alternatively, the beads can be reacted with chlorosulfonic acid to produce a sulfonyl chloride group. The sulfonyl chloride group is then hydrolyzed to the acid.

Presumably, the later reaction effects a more uniform placement of the ionogenic groups. This is because the chlorosulfonic acid reagent is dissolved in an organic solvent. The solvent swells the bead, allowing free access of the chlorosulfonic acid to the aromatic rings. Concentrated sulfuric acid is more polar. Sulfonation with this reagent occurs first on the bead surface and then moves progressively toward the center of the bead. Even though this product is not as homogeneous, resins prepared with concentrated sulfuric acid are more popular for ion chromatography. The $-SO_3^-$ anionic group that is produced is chemically bound to the resin and its movement is thus severely restricted. However, the H^+ counterion is free to move about and can be exchanged for another cation. When a solution of sodium chloride is brought into contact with a cation exchange resin in the hydrogen ion form, the following exchange reaction occurs:

Res—⟨benzene ring⟩—SO₃⁻H⁺ + Na⁺ ⇌ Res—⟨benzene ring⟩—SO₃⁻Na⁺ + H⁺ (3.4)

If this reaction goes essentially to completion, the resin is said to be in the sodium (ion) form.

Traditionally, anion-exchange resins have been made by a two-stage set of reactions (although other synthesis methods are now being used). The first step is a Friedel-Crafts reaction to attach the chloromethyl group to the benzene rings of styrene–divinylbenzene copolymer. Then the anion exchanger is formed by reaction of the chloromethylated resin with an amine. The most common type of strong base anion-exchange resin contains a quaternary ammonium functional group, which is obtained by alkylation with trimethylamine.

$$Res-\text{⟨ring⟩} \xrightarrow[\text{ZnCl}_2]{\text{ClCH}_2\text{OCH}_3} Res-\text{⟨ring⟩}-\text{CH}_2\text{Cl} \xrightarrow{\text{N(CH}_3)_3}$$

$$Res-\text{⟨ring⟩}-\text{CH}_2\text{N(CH}_3)_3{}^+\text{Cl}^-$$

(3.5)

In these resins only the anion is mobile and can be exchanged for another anion. Another common strong base anion exchanger is one that contains a hydroxyethyl group in place of a methyl group on the nitrogen. Weak-base anion exchangers are synthesized by reacting the chloromethylated resin with lower substituted amines or with ammonia. Weak-base anion exchange resins cannot function as ion exchangers unless the functional group is protonated:

$$\text{Resin-CH}_2\text{NH}_2 + \text{H}^+ + \text{NO}_3^- \rightarrow \text{Resin-CH}_2\text{NH}_3^+ \text{ NO}_3^- \qquad (3.6)$$

The protonation, of course, depends on the basicity of the functional group and the pH of the solution in which the resin beads are immersed.

3.2.5 Resin Capacity

Resin capacity is an extremely important parameter in ion chromatography. Details of the effect of capacity on the behavior of resins for ion chromatography are found in Chapter 5. Generally, resins of lower capacity will lower the eluent concentration needed to elute sample ions from the column.

The capacity of a resin is usually given in milliequivalents of exchangeable ion per gram of resin. In some cases it is expressed as milliequivalents per milliliter of resin. High-capacity commercial cation resins contain approximately one functional group per benzene ring and have an exchange capacity of around 4.5 mequiv/g. High-capacity anion-exchange resins are typically around 3.5 to 4.0 mequiv/g for a gel-type resin and around 2.5 mequiv/g for a macroporous resin.

In almost all cases, the bulk resins that are available commercially are high capacity. However, ion chromatography usually employs low-capacity ion exchangers. Capacities of these resins range from 0.01 to 0.2 mequiv/g. Many low-capacity resins are pellicular, that is, the ion-exchange groups are on or near the surface of the bead. However, macroporous resin substrates are porous and have a much higher surface area. Low-capacity ion exchangers made from these substrates have the functional groups distributed throughout the bead, although the exact relationship of the functional group within the resin matrix is unknown.

For anion chromatography, and especially for non-suppressed IC, it is necessary to have resins of low exchange capacity. Earlier work centered on functionalization of macroporous resins produced by Rohm and Haas.

The resin substrate studied most often in anion chromatography is XAD-1. The substrate has the lowest surface area (100 m^2/g) and the largest average pore diameter (205 Å) of the XAD series. The physical stability is excellent. XAD-1 can be converted to an anion exchanger very easily by chloromethylation, followed by amination with a tertiary amine.

Anion-exchange resins of variable but low exchange capacities are produced under mild conditions and short reaction times in the chloromethylation reaction. Conditions for the amination are chosen to convert as much of the chloromethyl group as possible to the quaternary ammonium chloride, although experience indicates that some of the chloromethyl remains unreacted.

A procedure devised by Barron and Fritz [1] uses concentrated hydrochloric acid and paraformaldehyde with a Lewis acid catalyst to chloromethylate the polymer. Exceptional control of the extent of chloromethylation is possible by adjusting the concentration of reagents, the reaction temperature, and the reaction time. Chloromethyl methyl ether is not generated in situ, except for possible traces, so the reaction is relatively safe to use. Following the chloromethylation step, amination is carried out by adding a large excess of 25 % trimethylamine in methanol or water and allowing the reaction to proceed overnight.

Depending on the conditions chosen, this procedure can produce anion exchangers with capacities from 0.005 to 0.16 mequiv/g when using XAD-1 as a substrate. This range includes the capacities most useful in ion chromatography.

3.3 Anion Exchangers

3.3.1 Poly(styrene–divinylbenzene) Backbone (PS-DVB)

The major commercial anion exchangers for ion chromatography are based on two substrate types: macroporous and microporous (or gel-type) materials. Microporous substrates are used primarily as supports for pellicular, latex-coated beads. This type of column is efficient and versatile. Many different column types with different selectivities are available. However, microporous substrates can be "spongy." Columns packed with this material may eventually have bed compression leading to reduced column performance.

Macroporous materials are rugged, and column beds made from this substrate are stable. However, some commercial materials show poor ion-exchange kinetics with certain eluents. Gjerde [2] described a macroporous anion-exchange resin that shows good IC separations with a variety of eluents including sodium carbonate/bicarbonate and sodium hydroxide. This resin, called the Sarasep AN1, is a highly cross-linked PS-DVB with 80 Å pores and 415 m^2/g surface area. The commercial material has an exchange capacity of 0.05 mequiv/g and an average particle size of 8 μm. The resin contains a quaternary ammonium functional group: dimethylethanol amine.

The substrate polymer beads used in this work are quite hard and do not swell and shrink when the solvent is changed. A measure of bead hardness is given by the swelling propensity value. This value is obtained by measuring the backpressure of a column with a tetrahydrofuran mobile phase and then with an aqueous phase. After correcting the eluent backpressure for viscosity, the swelling propensity, SP, is calculated by the following equation:

$$SP = \frac{THF_{pressure} - H_2O_{pressure}}{H_2O_{pressure}}$$

An SP value of 0 indicates non-swelling material. The substrate used in this work had an SP of 0.8. Silica-based substrates are hard and typically have SP values of 0.05. However, other PS-DVB resins often have SP values well above 1.0. Columns packed with the AN1 resin have given excellent separations of common anions using a variety of eluents.

Polymeric resins from Hamilton have been used extensively for anion chromatography. Their PS-DVB anex resins contain trimethylammonium groups. PRP X100 has 0.19 mequiv/g exchange capacity and PRP X110 has a capacity of 0.11 mequiv/g. Their RCX10 has a somewhat higher capacity, 0.35 mequiv/g. All of these have an average pore size of 100 Å. Good separations of anions have been obtained with a variety of eluents [3]. However, these resins are quite hydrophobic. For this reason, the eluents often contain ~7.5 % methanol or ~0.1 mM sodium thiocyanate to give better peak shapes.

A resin from Alltech, sold commercially as the Durasep A1, is another example of a PS-DVB anion exchanger [4]. A highly cross-linked backbone makes this material chemically and mechanically stable. It withstands organic solvents and is stable over a wide range of pH, temperature and pressure. The stationary phase is useful for the separation of inorganic anions by both suppressor and nonsuppressed chromatographic methods. It was shown that resolution and peak shape are improved by adding 5 to 15 % methanol to the eluent. In particular, methanol reduced or eliminated the hydrophobic interaction between nitrate and the resin, resulting in a symmetrical peak shape. However, addition of acetonitrile to the eluent seemed to increase the hydrophobicity of the column towards nitrate.

3.3.2 Polyacrylate Anion Exchangers

A wide variety of resins based on polyacrylate polymers has been produced for use in chromatography. A type known as HEMA, a macroporous copolymer of 2-hydroxyethyl methylmethacrylate and ethylene dimethacrylate, has been used extensively in ion chromatography. It is extensively cross-linked to produce a polymeric matrix with high chemical and physical stability. The structure of HEMA is shown in Fig. 3.2A. The tertiary carbonyl structure of pivalic acid is one of the most stable and least hydrolyzable esters known, which allows the HEMA stationary phase to be used with a variety of eluents in the pH range 2–12. The excess hydroxyl groups on the HEMA matrix also increase the hydrophilicity of this material, which will be shown later to result in improved peak shapes for polarizable anions. The strong-base anion exchanger of HEMA, shown in Fig. 3.2B, is prepared by treating the HEMA precursor with an aqueous solution of trimethylamine. The preparation procedures and the influence of different functional groups on sorbent selectivity were discussed by Vlacil and Vins [5].

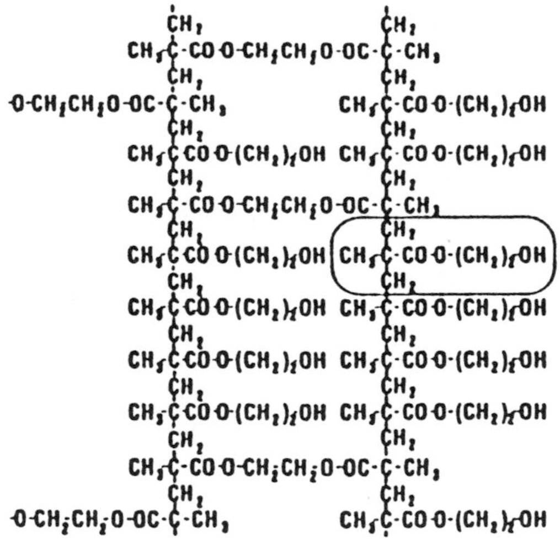

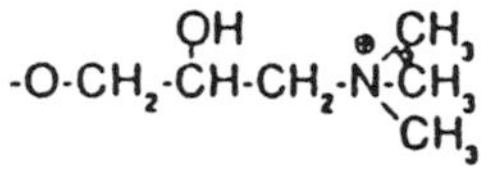

Figure 3.2. Structures of (A) HEMA and (B) strong-base anion exchanger of HEMA.

A HEMA-based anion exchanger developed by Alltech has been described as a universal stationary phase for the separation of a wide variety of anions [6,7]. This anion-exchange resin has been compared to agglomerated pellicular anion exchangers for the separation of anions by chemically suppressed IC. The HEMA-based columns exhibit higher capacities for all anions and particularly for weakly retained anions such as fluoride and formate, which were completely resolved. The HEMA columns could be used for both isocratic and gradient techniques.

3.3.3 Effect of Functional Group Structure on Selectivity

Anion exchange resins containing a benzyltrimethylammonium functional group are the most widely used type for anion chromatography. It was of interest to see how changing the chemical nature of the quaternary ammonium functional group might affect the selectivity of anion exchangers towards different anions. Barron and Fritz [8] prepared 13 different resins by reacting chloromethylate XAD-1 with different tertiary amines.

To ascertain the effect of functional group structure on selectivity, the various resins should have a very similar exchange capacity so that identical elution conditions could be used for each resin. To accomplish this, the relative reactivities of various amines with chloromethylated XAD-1 had to be determined. Then the degree of chloromethylation of XAD-1 could be adjusted so that the aminated resins would have similar capacities.

The retention times of 17 monovalent anions on resins with different functional groups but with almost identical exchange capacities (average: 0.027 mequiv/g) were compared with the use of a solution of a monovalent anion (sodium benzoate) as the eluent [8]. Relative retention times were calculated by dividing the measured retention times by that of chloride. Data for resins with various trialkylammonium groups are presented in Table 3.1.

Table 3.1. Relative retentions of anions on trialkylammonium resins [24].

Anion	TMA	TEtA	TPA	TBA	THA	TOA
Cl^-	1.0	1.0	1.0	1.0	1.0	1.0
F^-	0.66	0.70	0.69	0.71	0.68	0.69
Br^-	1.20	1.19	1.25	1.34	1.32	1.41
I^-	2.51	2.48	3.05	3.82	5.00	>5.0
$H_2PO_4^-$	0.84	0.85	0.84	0.85	0.83	0.83
NO_2^-	0.82	0.82	0.86	0.90	0.89	0.98
NO_3^-	1.30	1.32	1.38	1.54	1.63	1.72
Acetate	0.25	0.28	0.23	0.25	0.22	0.22
Formate	0.52	0.54	0.51	0.52	0.51	0.53
Lactate	0.44	0.49	0.46	0.47	0.45	0.49
Glycolate	0.45	0.48	0.44	0.45	0.43	0.45
Nicotinate	0.31	0.32	0.31	0.31	0.31	0.33
ClO_3^-	1.53	1.55	1.56	1.73	1.92	2.15
BrO_3^-	1.05	1.06	1.03	1.08	1.08	1.14
N_3^-	0.36	0.39	0.34	0.37	0.35	0.37
BF_4^-	2.70	2.58	3.41	4.34	>7.5	–
$CH_3SO_3^-$	1.00	1.06	1.01	1.06	1.01	1.07
t_{RCl}(min)	8.3	8.0	9.5	8.9	10.6	9.4

TMA = trimethylamine, TEtA = triethylamine, TPA = tripropylamine, TBA = tributylamine, THA = trihexylamine, and TOA = trioctylamine.

The data show that the relative retentions of the weak-acid anions are almost independent of the size of the alkyl groups. However, as the size of the R groups increase, large changes occur with the more polarizable anions such as nitrate, iodide, chlorate and BF_4^- ions.

The use of anion-exchange resins with different substituents offers a useful parameter for improving the separation of some anions. For example, Fig. 3.3 compares the separation of five anions on columns containing resins of approximately the same capacity but with increasingly larger alkyl groups on the quaternary nitrogen. Identical elution conditions were used. The TMA resin column gave poor resolution of bromide and nitrate. The resolution was improved on the TPA column, and a baseline separation was obtained on the THA column.

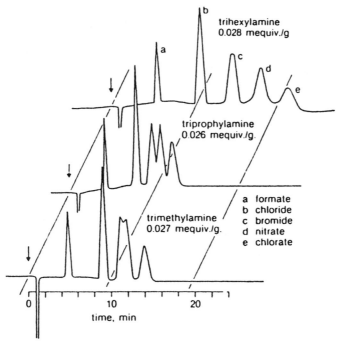

Figure 3.3. Separation of five anions on three different resins of similar capacity. The resins were packed in a 500 × 2.0 mm i.d. column and a solution of 0.0001 M benzoic acid was used as the eluent at a flow rate of 0.93 mL/min (From Ref. [8], with permission].

Most of the monovalent anions exhibit only small changes in their relative retention on resins containing one, two, or three hydroxyethyl groups, compared to the trimethylamine resins (Table 3.1). However, when a stronger eluent was used (phthalate, 2- instead of benzoate, 1-) and several divalent anions were examined, the changes resulting from hydroxyethyl groups became more apparent [9].

The data in Table 3.2 show that MDEA considerably lowers the relative retention of nitrate, chlorate, iodide and thiocyanate compared to the TMA resin. It also shows that with the phthalate eluent the tributyl resin (TBA) has longer retention times for nitrate and much longer for iodide, but shorter for sulfate and thiosulfate, all compared to the TMA material.

Table 3.2. Relative retentions of ions on three anion exchangers of differing polarity. Phthalate eluent (0.4 mM), pH 5.0.

Ion	MDEA*	TMA**	TBA***
Cl^-	1.0	1.0	1.0
NO_3^-	1.47	1.79	2.77
ClO_3^-	1.78	2.42	3.18
I^-	3.81	6.16	13.9
SCN^-	7.50	14.5	–
ClO_4^-	9.12	–	–
SO_4^{2-}	7.31	7.29	6.36
$S_2O_3^{2-}$	15.75	16.6	9.23
$C_2O_4^{2-}$	7.41	7.18	6.30
MoO_4^{2-}	9.50	9.92	8.72
WO_4^{2-}	–	–	–
t_{RCl^-} (min)	3.20	3.80	3.90

Values are expressed at t_{Ranion}/t_{RCl^-}.
* Capacity 0.090 mequiv/g.
** Capacity 0.092 mequiv/g.
*** Capacity 0.096 mequiv/g.

Virtually all anion exchange separations have been carried out with resins containing a single nitrogen atom in each exchange group, either a quaternary ammonium group ($-N^+R_3$) or a protonated amine group ($-N^+HR_3$). A novel resin has been described, containing three nitrogen atoms in each functional group [10]. A chloromethylated PS-DVB resin was reacted with diethylenetriamine to give a functional group of structure a, or b, or a mixture of the two.

a. $CH_2 - NH - CH_2 - CH_2 - NH - CH_2 - CH_2 - NH_2$

b. $CH_2 - N$ with $CH_2 - CH_2 - NH_2$ and $CH_2 - CH_2 - NH_2$

By varying the pH at which the resin is used in an IC column, one, two or three of the N atoms can be protonated giving a net charge of 1+, 2+ or 3+ for each functional group.

The retention times of sample anions become longer as the operating pH becomes more acidic and the net positive charge on the ion exchanger increases. Figure 3.4 plots the retention factor as a function of eluent pH for several sample anions. Thiocyanate and molybdate are very strongly retained, even at moderately acidic pH values.

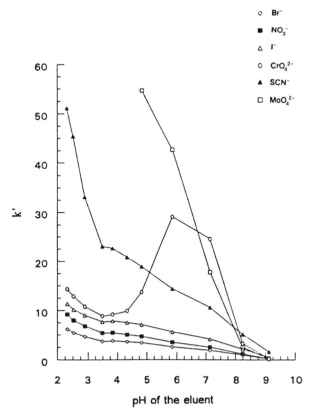

Figure 3.4. Eluent pH vs. anion capacity factor, k'. Conditions: 100×4.6 mm (multicharge, weak base, anion exchanger) column, 15 mM sodium perchlorate eluent, UV detection at 200 nm (from Ref. [10], with permission).

Several anions were separated chromatographically at pH 7.5–7.7, with different salts in the mobile phase, and with direct UV detection at 200 nm. The results in Table 3.3 show that perchlorate is a significantly better eluting anion than chloride is. However, sulfate and hydrogen phosphate both give even shorter retention times by virtue of their 2– charge.

Table 3.3. Retention times (min) of several anions with different eluents.

Anion	$NaClO_4$[a]	$NaSO_4$[a]	$NaCl$[a]	Na_2HPO_4[a]
Bromide	3.45	2.15	4.51	1.98
Nitrate	3.98	2.75	5.55	2.54
Iodide	6.25	7.01	12.4	5.95
Thiocyanate	13.7	ND[b]	25.6	17.4

[a] Each eluent: 5.0 mM at pH 7.5–7.7.
[b] ND = not detected.

3.3.4 Effect of Spacer Arm Length

Polymeric anion exchangers are normally prepared by chloromethylation of the benzene ring, followed by reaction with a tertiary amine to give a quaternary ammonium group. Thus the N^+ is connected to the benzene ring by a single $-CH_2$ group. Suppose the N^+ was connected to the benzene ring by a longer series of $-CH_2$ groups, sometimes called the spacer arm. What effect would a longer spacer arm have on ion exchange selectivity?

Warth and Fritz synthesized a series of resins with spacer arms of varying lengths [11]. The benzene ring of a macroporous resin (Rohm & Haas XAD-1) was reacted under controlled conditions with a bromoalkene with CF_3SO_3H as catalyst, and was then quaternized by reaction with trimethylamine. This gave a resin of the following structure:

$$Res \!-\!\!\left\langle\bigcirc\right\rangle\!\!- CH \underset{\underset{CH_3}{|}}{} (CH_2)_n\ N^+(CH_3)_3Br^-$$

Conditions were adjusted so that all of the resins had almost identical exchange capacities. The spacer arm length varied from one to six methylene groups.

Chromatographic retention times of a number of anions were compared with 6.0 mM nicotinic acid and 2.0 mM phthalate (pH 6) as the eluents. In many cases, the length of spacer arm had very little effect on the relative retention times (relative to Cl^-). However, the selected data in Table 3.4 show that the relative retention times of bromide, nitrate, chlorate and iodide decreased while that of sulfate increased slightly. Longer spacer arms would of course reduce any influence the benzene ring might have on ion exchange retention.

Table 3.4. Adjusted retention times as a function of spacer arm length of anion exchange resins.

| | Spacer arm length | | | | |
Anion	C1	C2	C3	C4	C6
Chloride	1.0	1.0	1.0	1.0	1.0
Nitrite	1.8	1.5	1.5	1.4	1.3
Bromide	2.5	2.1	2.0	1.8	1.9
Nitrate	3.6	3.1	2.5	2.4	2.6
Chlorate	7.1	4.5	3.9	3.7	4.1
Sulfate	7.4	8.3	8.5	8.5	8.5
Iodide	17.6	12.2	8.9	8.6	11.2

3.3.5 Quaternary Phosphonium Resins

An anion exchange resin for IC has been prepared by reaction of a chloromethyl PS–DVB resin with tributylphosphine (TBP).

$$
\text{Res} - \langle\!\!\bigcirc\!\!\rangle - CH_2Cl + (C_4H_9)_3P \rightarrow \text{Res} - \langle\!\!\bigcirc\!\!\rangle - CH_2P^+(C_4H_9)_3Cl^- \tag{3.7}
$$

The TBP resin is quite stable and is suitable for anion chromatography. Warth, Cooper and Fritz [12] compared the retention times of several anions relative to chloride using columns packed with quaternary ammonium anion exchangers of the conventional trimethyl type (TMA) and tributylamine (TBA). The selected results in Table 3.5 show that bromide, nitrate, chlorate and iodide are retained more strongly by the TBP resin.

Table 3.5 shows that nitrate elutes much later from a TBP column than either chloride or sulfate. The chromatographic separation of traces of chloride from 200 times as much nitrate was possible [12]. A U.S. Patent was issued for the selective removal of nitrate from drinking water [13].

Table 3.5. Relative retention times (min) of selected anions on columns prepared from trimethylamine (TMA), tributylamine (TBA) and tributylphosphine (TBP). Eluent: 2.2 mM sodium phthalate at pH 6.0.

Anion	TMA	TBA	TBP
Chloride	1.0	1.0	1.0
Nitrite	1.5	1.6	1.6
Bromide	1.9	2.3	2.7
Sulfate	2.1	2.4	2.3
Nitrate	2.5	3.7	4.3
Chlorate	4.3	4.3	5.3
Iodide	9.6	17.9	29.1

3.3.6 Latex Agglomerated Ion Exchangers

Pellicular materials in which the stationary phase is a layer on the outside perimeter of a spherical substrate have been used frequently in liquid chromatography. The relatively thin layer ensures a rapid equilibrium between the mobile and stationary phases even when the column packing has a relatively large particle size. In their original work on ion chromatography, Small et al. [14] used a surface-sulfonated material as a pellicular cation exchanger. Such materials are easy to prepare because sulfonation of a polymer containing benzene rings proceeds from the outside in. Sulfonation for a short period under mild conditions will insert sulfonic acid groups only on or near the outside surface of a spherical resin.

It was soon discovered that efficient anion exchange resins could be prepared by coating the outside of a surface-sulfonated polymer with a layer of latex particles functionalized with quaternary ammonium groups. The first commercial anion exchangers for ion chromatography (Dionex ASI) consisted of 0.15 µm latex particles coated onto a 25 µm sulfonated substrate. A two-dimensional diagram of this coating is shown in Fig. 3.5A. The positively charged latex particles are firmly held by electrostatic attraction as shown in Fig. 3.5B. Each latex particle has several quaternary N^+ groups, so the coated substrate will have many quaternary groups available for ion exchange. Latex agglomerated resins are very stable chemically. Even 4 M sodium hydroxide is unable to cleave the ionic bond between the substrate and the latex bead.

a)

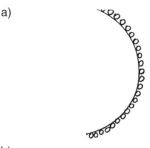

b)

Substrate - SO_3^- $\underset{R_3}{N^+}$ - Latex

Figure 3.5. a)
b) Substrate–$SO_3^-N^+R_3$–Latex

Several advantages have been cited for latex coated anion exchangers [15]:
- The substrate provides mechanical stability and gives a moderate backpressure.
- The small size of the latex beads and their location on the outer surface of the substrate ensure fast exchange processes and thus a high chromatographic efficiency.
- Swelling and shrinkage are minimal.

The properties of latex resins can be varied by manipulation of several parameters. Hydrophobic attraction of the exchanger for some anions can be altered by varying the type and cross-linking of the polymeric substrate. The ion-exchange capacity is determined by the substrate particle size, the size of the latex beads, and the degree of latex coverage on the substrate surface. Selectivity for various anions is governed mainly by the type of functional groups attached to the latex bead and by the degree of latex cross-linking.

Over the years, Dionex has developed a wide variety of latex agglomerated resins to meet various needs in IC. A review of these developments is given in a book by Weiss [16]. Properties of latex columns are summarized in Table 3.6.

A method has been described for preparation of latex-coated anion exchange resins that does not involve sulfonation of the substrate [17]. A suspension of quarternized latex beads in water containing 0.01 to 0.10 M sodium chloride is used to coat an unfunctionalized polymeric substrate. Once coated, the latex sticks tightly and is not washed off by aqueous solutions. The latex can be removed by washing with pure organic solvents and may then be recoated. The exchange capacity may be varied by changing the concentration of sodium chloride or latex in the coating solution. Various resin substrates were coated with exchange capacities ranging from 5 to 400 µequiv/g. Columns packed with these latex exchangers gave unusually efficient separations of sample anions.

Table 3.6. Retention factors (k) for different latex functional groups. Eluent: 5 mM sodium carbonate.

Column	F^-	Cl^-	Br^-	NO_3^-	ClO_3^-	SO_4^{2-}	HPO_4^{2-}
MDEA	0.09	0.54	2.2	2.6	2.6	5.2	3.3
DMEA	0.18	1.0	3.7	4.4	4.2	5.3	6.4
TMA	0.16	1.2	4.5	5.4	5.3	3.3	7.6
TEA	0.13	1.9	9.6	16.8	9.6	3.4	5.6

3.3.7 Effect of Latex Functional Group on Selectivity

Slingsby and Pohl [18] investigated the effect of varying the structure of the quaternary ammonium group on the latex while keeping the percentage of cross-linking and the polymeric backbone structure constant. They estimated that each 5 µm spherical substrate particle was coated with approximately 28 000 quaternized latex beads. Retention factors (k), corrected for column capacity, were measured for four different columns. The latex functional group in column 1 was methyldiethanolamine (MDEA), column 2 was dimethylethanolamine (DMEA), column 3 was trimethylamine (TMA), and column 4 was triethylamine (TEA). The eluent in Table 3.6 was 5 mM sodium carbonate; the eluent in Table 3.7 was 100 mM sodium hydroxide.

Table 3.7. Retention factors (k) for different latex functional groups. Eluent: 100 mM sodium hydroxide.

Column	F^-	Cl^-	Br^-	NO_3^-	ClO_3^-	SO_4^{2-}	HPO_4^{2-}
MDEA	0.06	0.24	0.92	1.1	1.0	0.20	0.31
DMEA	0.14	1.1	4.5	5.0	4.9	3.0	6.7
TMA	0.30	4.4	19.2	22.5	21.2	51.4	>100
TEA	0.30	5.8	26.1	55.8	24.9	24.9	>100

Two important conclusions may be drawn from these tables. One is that the hydroxide ion is a much weaker eluent for anion chromatography than carbonate. The second is that the eluting power of sodium hydroxide is enhanced considerably by using latexes with one or two hydroxyethyl groups instead of those containing only alkyl groups.

Surprisingly good ion exchangers can be prepared simply by coating an ordinary HPLC material, such as C18 silica or PS-DVB, with a hydrophobic molecule containing an ionic functional group. Table 3.8 lists a number of coating materials that have been used [19].

The actual coating may be done statically by equilibrating the particles of bonded-phase silica or polymeric resin with a solution of the coating material in methanol–

water or acetonitrile–water. A column of the type used for HPLC may be coated dynamically by passing an excess of the coating solution continuously through the packed column. As a specific example, a good anion-exchange column may be prepared by coating a C18 silica column with a hydrophobic quaternary ammonium salt, such as tridodecylmethylammonium iodide [19]. An aqueous phosphate solution (pH 6.4) was used as the mobile phase. Separation of anions on a PS-DVB (Hamilton PRP-1) coated column required an aqueous mobile phase containing 15 % acetonitrile. The acetonitrile was needed to make good surface contact with the very hydrophobic PRP-1.

Table 3.8. Reagents for coated ion-exchange resins [5,6].

	Carbon number	Reagent salt
Cation exchange:		
	6	1-hexanesulfonate
	8	1-octanesulfonate
	12	1-dodecylsulfate
	20	1-eicosylsulfate
Anion exchange:		
	8	tetraethylammonium
	16	tetrabutylammonium
	21	cetylpyridinium
	25	trioctylmethylammonium
	32	tetraoctylammonium
	37	tridodecylmethylammonium

Other authors have used coated columns successfully for ion chromatography. Polymer-based packing or a C18 silica packing coated with methyl green or ethyl violet [4] has been used to coat substrates for the separation of anions. Several dyes have been used to coat substrates for separation of various metal ions [5].

A comprehensive study of coated anion-exchange resins provides some valuable insights [20]. Of several coating materials tested, cetyl pyridinium chloride (CPI) was found to be the most satisfactory. Static coating was employed with $1-3 \times 10^{-4}$ M CPCI in 3 % (v/v) acetonitrile–97 % water. Coating of a polyacrylate resin (Rohm & Haas XAD-8) gave sharper chromatographic peaks for anions than the more hydrophobic PS-DVB resin (Hamilton PRP-1).

The coating thickness of CPCI on the polyacrylate resin, and hence the ion exchange capacity, could be varied by adjusting the static coating conditions. At very low capacities (7.5 to 12.1 μequiv/g) linear plots of adjusted retention time versus exchange capacity were obtained for test ions such as fluoride, chloride, bromide and nitrate.

Experiments with several different hydrophobic quaternary ammonium salts as coating reagents showed that the adjusted retention times of test anions relative to t' for chloride varied somewhat with the chemical structure of the coating [20]. However, it was found that the relative adjusted retention times also varied considerably with the type of resin substrate. This is demonstrated in Table 3.9 where relative adjusted retention times are compared: (1) on a conventional anion exchanger with benzyltrimethyl ammonium functional groups, (2) on cetylpyridinium chloride coated on polystyrene-DVB XAD-1 resin, and (3) on cetylpyridinum coated on polyacrylate XAD-8 resin. These data show that polarizable anions such bromide, nitrate and especially iodide have much longer retention times on the coated polyacrylate resin than on the coated PS-DVB resin. On the other hand, the relative retention time of sulfate is 2.4 times higher on the PS-DVB coated resin than on the polyacrylate material.

Table 3.9. Adjusted retention times for anions relative to chloride. Capacity of trimethylamine XAD-1 was 0.027 mequiv/g; theoretical capacity of cetylpyridinium coated resins was 0.050 mequiv/g. Eluent was 0.2 mM tetrabutylammonium phthalate, pH 6.5. Conductivity detection.

Anion:	Chemically-coated bonded exchanger:	Coated Exchanger:		
	Trimethylamine, XAD-1	CPCl, XAD-1	CPCl, XAD-8	TOACl, XAD-1
Chloride	1.00	1.00	1.00	1.00
Nitrite	1.18	1.11	1.49	1.12
Methylsulfonate	1.18	1.01	1.03	1.13
Bromate	1.23	1.45	1.23	1.29
Bromide	1.38	1.49	2.05	1.30
Nitrate	1.54	2.37	3.67	1.44
Chloroacetate	1.54	1.58	1.33	1.45
Chlorate	2.10	3.20	5.08	1.84
Iodide	3.49	4.35	12.7	2.12
Chromate	7.00	6.42	5.44	7.72
Thiocyanate	8.95	–	9.72	–
Sulfate	10.3	9.00	3.79	8.06
Thiosulfate	–	17.8	6.03	10.7
t'_{Cl} (min)	1.17	2.73	1.17	4.62

3.3.8 Silica-Based Anion Exchangers

Ion exchangers are available in which an organic material containing a quaternary ammonium functional group is chemically bonded to porous silica spheres. This results in a thin layer of ion-exchange material on the silica surfaces. Vydac IC 302 [22] is one such resin. This is a spherical silica of high mechanical strength with a particle diameter of approximately 15 μm. The particles have a surface area of 86 m²/g and an average pore diameter of 330 Å.

Compared to organic polymers, silica-based ion exchangers have the advantages of higher chromatographic efficiency and greater mechanical stability. In general, no problems due to swelling or shrinking are encountered, even if an organic solvent is added to the eluent. A disadvantage of silica materials is their limited stability at lower pH values and especially in alkaline solutions. A fairly narrow pH range of 2 to 8 is recommended.

There are two distinct types of silica resins. Totally porous resins have a quaternary ammonium functional group chemically attached. Their particle size is in the range of 3 μm to 10 μm with typical exchange capacities of 0.1 to 0.3 mequiv/g. Pellicular materials have a larger particle size and are covered with a thin layer of a polymer with quaternary ammonium groups. For example, Zipax SAX is covered with a layer of lauryl methacrylate 1 μm to 3 μm thick.

Silica-based anion exchangers are available from a number of manufacturers. Typical trade names include Vydac (Separations Group, Hesperia, CA, U.S.A.), Wescan (Alltech, Deerfield, IL, U.S.A.), TSK Gel (Toyo Soda, Tokyo, Japan) and Nucleosil (Machery & Nagel, Düren, Germany).

3.3.9 Alumina Materials

Pietrzyk and coworkers [21] showed that hydrated alumina can function as a low-capacity anion exchanger. The isoelectric pH for alumina is about 7.5 and its ion-exchange capacity (about 2 mequiv/g maximum) is dependent on the sample ion, the eluent pH, and the pretreatment of the alumina. The alumina must be first hydrated and then treated with either an acid or base. Anion-exchange capacity increases as the eluent pH is made more acidic. The changeover from an anion exchanger to a cation exchanger is gradual and occurs in the vicinity of eluent pH 7.5.

In anion chromatography the elution order of common anions is almost the reverse of conventional resins. Fluoride and phosphate are so strongly held by the alumina that elution is often impossible.

3.4 Cation Exchangers

3.4.1 Polymeric Resins

3.4.1.1 Sulfonated Resins

Most of the cation exchangers used in IC fall into two major categories: sulfonated resins, sometimes called "strong-acid" exchangers, and resins with carboxylic acid groups, sometimes called "weak-acid" exchangers. The ion exchangers used in IC have a much lower exchange capacity than those intended for commercial applications such as the removal of calcium and magnesium ions from hard water.

Low-capacity cation-exchange resins are obtained by superficial sulfonation of styrene–divinylbenzene copolymer beads. The resin beads are treated with concentrated sulfuric acid and a thin layer of sulfonic acid groups is formed on the surface. The final capacity of the resin is related to the thickness of the layer and is dependent on the type of resin, the bead diameter, and the temperature and time of contact with the sulfuric acid. Typical capacities range from 0.005 to 0.1 mequiv/g.

It can be easily appreciated that, compared to a conventional cation-exchange resin, the diffusion path length is reduced because the unreacted, hydrophobic resin core restricts analyte cations to the resin surface. This results in faster mass transfer of the cations and consequently in improved separations. Also, because of the rigidity of the resin core, there is less tendency for the bead to compress. This means that higher flow rates (at relatively low back pressures) can be used than would be possible with conventional resins. Superficially functionalized resins are stable over the pH range of 1 to 14 and swelling problems are minimal. The selectivity of the superficial cation-exchange resins for ions is similar to that observed for conventional resins.

For chromatographic purposes it is necessary to control the conditions for sulfonation so that the polymer beads are sulfonated evenly and that the exchange capacity is very reproducible. It is also desirable to be able to increase or decrease the resin capacity in a predictable fashion.

Pepper [22] proposed that the sulfonation of neutral poly(styrene–divinylbenzene) copolymer beads proceeded in a "layer-by-layer" fashion and could be stopped at any particular depth, giving a partially sulfonated bead with a low capacity and a known depth of functionality. From this concept Pepper [23] prepared the first superficially sulfonated polymer beads. Parrish [24] also prepared surface-sulfonated beads and gave a brief example of their utility.

Fritz and Story [25,26] prepared several low-capacity sulfonated resins for the chromatographic separation of various metal cations. Macroporous resins were used and their selectivity was somewhat different from that of conventional gel resins.

The surface sulfonation of poly(styrene–divinylbenzene) beads with cross-linking ranging from 0.5 to 8 % divinylbenzene was studied by Small [27,28]. Examination of the sulfonation depth of a resin bead in terms of optimum separation of several inorganic cations was subsequently studied by Stevens and Small [29]. These resins were used for the chromatographic separation of simple cations.

Low-capacity resins have been prepared specifically for use in ion chromatography by Fritz et al. [30]. The resin used in the separations was a 3:2 blend of neutral resin with low-capacity sulfonated resin giving a final capacity of about 8 µeq/g. Very good efficiency was demonstrated for separation of the alkali cations and alkaline earths.

Papanu et al. [31] fabricated a cation-exchange resin by agglomerating a sulfonated latex onto larger beads of a low-capacity anion-exchange resin. Battaerd [32,33] prepared a superficially sulfonated bead by graft-polymerization of a sulfonated olefin onto a polyolefin core. Kirkland [34] impregnated a porous fluoropolymer bead with a sulfonated fluoropolymer, giving a pellicular type of ion exchanger. Horvath et al. [35] polymerized a coating of polystyrene–DVB onto glass beads and formed a cation exchanger by sulfonation. Several authors have prepared cation exchangers by introduction of sulfonated organic group onto the surface of silica supports [36–38].

Sevenich and Fritz [39] examined the sulfonation of resins for use in IC in some detail. Spherical microporous resin beads of 4 %, 6.5 %, and 12 % cross-linking were selected for their study. Initially, sulfonation of 4 % cross-linked resin beads under identical conditions gave poor reproducibility. This was partly due to agglomeration of the small resin beads. Dispersion of the resin beads and even initial wetting of the surface by sulfuric acid seemed to be major problems in achieving reproducible sulfonation. The procedure finally developed involved mechanical and ultrasonic dispersion of the particles in methanol, passing the resin slurry through a small sieve, then removal of as much methanol as possible by suction filtration. The resin was then sul-fonated. The small amount of methanol present when the resin is added to the hot sulfuric acid immediately volatilizes and is swept out in the inert gas stream. This is confirmed by a short burst of vapor issuing from the reaction vessel when the resin is added. The capacity is approximately linear with reaction time and the reproducibility was ±5 %.

Sulfonation of gel beads of 4 %, 6.5 % and 12 % cross-linking was compared. Data for sulfonation of each for 30, 60 and 90 min are given in Table 3.10. These results show that resins of low capacity can be obtained in all cases but that the reaction is better controlled with resin beads of higher cross-linking.

Table 3.10. Capacity[a] (μequiv/g) of surface-sulfonated resins with respect to reaction time for 4, 6.5, and 12 % divinylbenzene cross-linked gel resins at 80 °C.

	Capacity (μequiv/g)		
Reaction time (min):	30	60	90
Cross-linking (% DVB):			
4	33	50	88
6.5	12	22	47
12	6	14	25

[a] Each value is the average of three individual results. The average deviation was approximately ±5 %.

The location of sulfonate groups in the resin bead was visualized by completely replacing H^+ with UO_2^{2+} as the counter ion and obtaining transmission electron micrographs on thin slices of the resin bead. Because uranium is a very heavy metal, the uranyl ions have higher stopping power for electrons and appear as a darker area on the micrograph. A dark outline around the resin slice, indicated that the uranyl ions (and hence the sulfonate groups) are located in a thin zone (approximately 200 Å) at the outer perimeter of the resin bead.

Knowing the resin capacity and the estimated thickness of the sulfonated layer, a simple calculation shows nearly complete sulfonation within the sulfonated layer. By "complete," we mean that there is approximately one sulfonate group for each ben-zene ring of the polymer. The density of sulfonate groups in this layer is similar to that in the entire bead of a typical high-capacity cation exchange resin.

Columns packed with the 12 % cross-linked resin, 6.1 mequiv/g exchange capacity gave good separations of metal cations. Using 8 different concentrations of perchloric acid eluents, from 0.10 to 1.00 M, retention times for 36 metal ions were measured [40]. The selectivity data obtained are given in Tables 5.4 to 5.9.

3.4.1.2 Weak-Acid Cation Exchangers

Owing to their differences in selectivity, it is often difficult to find conditions for separation of cations of different positive charge on a sulfonated resin column. Elu-ents that provide good separation of monovalent cations are too weak to elute divalent cations in a reasonable time. There is now a trend to use weak-acid cation exchange columns. These materials contain carboxylic acid functional groups, or in some cases mixed carboxylic acid and phosphonic acid groups. At more acidic pH values these groups are gradually converted from the ionic to the molecular form and thus their ability to retain sample cations is diminished. By adjusting the operating pH to an appropriate value it becomes possible to separate a wider variety of cations in a single run. One company advertises their chromatographic columns of this general type as a Universal Cation Column.

The properties and performance of a commercial weak-acid resin column (Dionex CS12) have been described [41]. The substrate is a highly cross-linked, macroporous ethylvinylbenzene–divinylbenzene polymer with a bead diameter of 8 μm, a pore size of 6 nm, and a specific surface area of 300 m^2/g. In a second step, this substrate was grafted with another polymer containing carboxylate groups. The exchange capacity is listed as 2.8 mequiv/column for a 250 mm × 4 mm i.d. column. With this column, simple eluents such as hydrochloric or methanesulfonic acid can be used to separate mono- and divalent cations rapidly and efficiently under isocratic conditions.

Morris and Fritz [42] described the preparation and chromatographic applications of two weak-acid resins that are easily synthesized, and carry the exchange group on the cross-linking benzene ring of the resin or on a short spacer arm from the ring. The first resin was prepared by reaction of a cross-linked polystyrene resin with succinic anhydride in a Friedel–Crafts reaction, aluminum chloride catalyst. The carboxyl groups are connected to the resin benzene rings by a three-carbon atom spacer arm: $-COCH_2CH_2CO_2H$. The second cation exchanger was prepared by reaction of the resin with phenylchloroformate to give a phenyl ester attached to the resin benzene rings: $-CO-OC_6H_5$. The ester groups were then hydrolyzed by refluxing for 1 hour in a sodium hydroxide–ethanol solution to give the sodium salts of the carboxylate. The exchange capacity of resin I was 0.60 mequiv/g and that of resin II was 0.39 mequiv/g. Resin II in particular gave excellent separations of divalent metal cations with a complexing eluent.

3.4.2 Pellicular Resins

The Dionex Co. has developed a number of latex-coated cation exchange materials. Their anion-exchange resins have a surface layer of quaternary ammonium latex on a surface-sulfonated substrate. By adding a second coating layer of sulfonated latex beads, the outer layer of the resin consists of latex beads with exposed sulfonate groups. These groups undergo cation exchange with sample and eluent cations in IC. A schematic representation for these cation exchangers is:

$$\text{Resin--SO}_3^- \; \overparen{(\text{N}^+\text{R}_3 \text{ latex N}^+\text{R}_3)} \; \overparen{(\text{SO}_3^-\text{-latex-SO}_3^-)} \qquad (3.8)$$

A diagram of such a resin (Ion Pac CS3) indicates that the sulfonated latex beads are of larger diameter than the quaternized beads (Fig. 3.6).

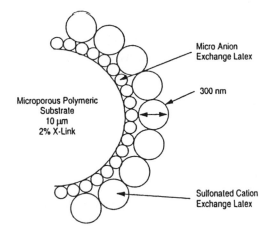

Figure 3.6. Schematic representation of IonPac CS3, a latex-coated pellicular strong-acid cation exchanger (Courtesy Dionex Corp).

The major Dionex cation exchange resins are listed in Table 3.11. Those through CS11 appear to be coated with latex or other very small polymer particles. However, the CS12 resin has a thin layer of a polymer with carboxyl groups grated to the surface of the substrate particles. This and their other more recent cation exchangers would seem to indicate a trend away from latex-coated cation exchangers. It will also be noted that all of the more recent resins contain more weakly acidic exchange groups instead of the strongly acidic sulfonic acid groups.

3.4.3 Silica-Based Cation Exchangers

Several of the earlier cation exchangers contained groups such as $-(CH_2)_3C_6H_4SO_3^-$ attached to spherical silica particles, but these no longer find much use in IC.

Considerable interest has been shown in a novel cation exchanger first developed by Schomburg et. al. [43]. The material consists of a silica substrate of very uniform particle size coated with a poly(butadiene-maleic acid) resin which serves as the cation-exchange moiety. This material, which is now commercially available, gives good separations of both monovalent and divalent metal ions in a single run. Ordinary eluents such as hydrochloric or methanesulfonic acid, or complexing eluents may be used [44,45].

A novel silica-based cation exchanger is functionalized with a combination of carboxylate and a crown ether functionalities [46]. This stationary phase is more selective toward ammonium, potassium and low-molecular weight amines. Potassium is eluted after ammonium, magnesium and calcium.

3.5 Chelating Ion-Exchange Resins

The selectivity of ordinary cation-exchange resins for various metal ions is somewhat limited. However, if a suitable chelating functional group is built into a polymeric resin, it often is possible to take up only a small group of metal ions. Other chelating resins may complex a larger group of metal ions, but selectivity is attained through pH control. Chelating resins also are valuable in sorbing a desired metal ion (or small group of metal ions) from solutions containing a very high concentration of a non-complexed metal salt. Frequently the selectivity of a chelating resin is so great that a very short column can be used to retain the desired metals.

Although a great many chelating resins have been described, only a few have sufficiently fast rates of metal ion equilibrium and chromatographic efficiency for practical separations. Several chelating functional groups for resins are listed in Table 7.8. Chromatographic separations using chelating resins are discussed in Section 7.5.

Table 3.11. Dionex cation-exchange resins.

Ion Pac #	Brief description
CS3	10 μm substrate coated with 50 nm aminated latex and then with 250 nm sulfonated latex, 5 % cross-linked
CS5A	Bifunctional pellicular, both catex and anex functions. Highly cross-linked
CS10	Aminated colloidal polymer layer covalently bound to substrate of high cross-linking. 100 % solvent stability, moderate capacity, hydrophilic surface
CS11	Similar to CS10 but higher capacity
CS12	Has a thin polymer layer with carboxyl groups grafted to a polymeric substrate
CS12A	Hydrophilic surface. Contains both carboxyl and phosphonic acid groups
CS14	Ethylvinylbenzene–DVB copolymer, 8 μm, with carboxyl functional groups. Solvent-compatible
CS15	Contains carboxyl, phosphonate and crown ether groups. Good separation of NH_4^+ and K^+ ethylvinylbenzene–DVB, 8.5 μm particles

References

[1] R. E. Barron and J. S. Fritz, Reproducible preparation of low-capacity anion exchange resins, *Reactive Polymers*, 1, 215, 1983.
[2] D. T. Gjerde, New macroporous stationary phase for the separation of anions. Advances in Chromatography, vol. 2, p 169. Century International, Medfield, MA 1990.
[3] Hamilton Company, Polymeric HPLC columns. Hamilton Co., Reno, NV, U.S.A.
[4] L. M. Nair, B. R. Kildew and R. Saari-Nordhaus, Enhancing the anion separations on a polydivinylbenzene-based anion stationary phase, *J. Chromatogr. A*, 739, 99, 1996.
[5] F. Vlacil and I. Vins, Modified hydroxyethyl methacrylate copolymers as sorbents for ion chromatography, *J. Chromatogr.*, 391, 133, 1987.
[6] R. Saari-Nordhaus, I. K. Henderson and J. M. Anderson, Jr., Universal stationary phase for the separation of anions on suppressor-based and single-column ion chromatographic systems, *J. Chromatogr.*, 546, 89, 1991.
[7] R. Saari-Nordhaus and J. M. Anderson, Jr., Applications of an alternative stationary phase for the separation of anions by chemically suppressed ion chromatography, *J. Chromatogr.*, 602, 15, 1992.

[8] R. E. Barron and J. S. Fritz, Effect of functional group structure on the selectivity of low-capacity anion-exchangers for monovalent anions, *J. Chromatogr.*, 284, 13, 1984.

[9] R. E. Barron and J. S. Fritz, Effect of functional group structure and exchange capacity on the selectivity of anion exchangers for divalent anions, *J. Chromatogr.*, 316, 201, 1984.

[10] J. Li and J. S. Fritz, Novel polymeric resins for anion-exchange chromatography, *J. Chromatogr.*, 793, 231, 1998.

[11] L. M. Warth and J. S. Fritz, Effect of length of alkyl linkage on selectivity of anion exchange resins, *J. Chromatogr. Sci*, 26, 630, 1988.

[12] L. M. Warth, R. S. Cooper and J. S. Fritz, Low-capacity quaternary phosphonium resins for anion chromatography, *J. Chromatogr.*, 479, 401, 1989.

[13] J. E. Lockridge and J. S. Fritz, Decontamination of water using a nitrate-selective ion-exchange resin. U.S. Patent 4,944,878, July 31, 1990.

[14] H. Small, T. S. Stevens and W. G. Bauman, Novel ion-exchange chromatographic method using conductimetric detection, *Anal. Chem.*, 47, 1801, 1975.

[15] J. Weiss, Ion Chromatography, 2nd Ed., p 43, VCH, Weinheim, Germany, 1995.

[16] J. Weiss, Ion Chromatography, 2nd Ed, p 43–55, VCH, Weinheim, Germany, 1995.

[17] L. M. Warth, J. S. Fritz, and J. O. Naples, Preparation and use of latex-coated resins for anion chromatography, *J. Chromatogr.*, 462, 165, 1989.

[18] R. W. Slingsby and C. A. Pohl, Anion-exchange selectivity in latex-based columns for ion chromatography, *J. Chromatogr.*, 458, 241, 1988.

[19] R. M. Cassidy and S. Elchuk, Dynamically coated columns for the separation of metal ions and anions by ion chromatography, *Anal. Chem.*, 54, 1558, 1982.

[20] D. L. Duval and J. S. Fritz, Coated anion-exchange resins for ion chromatography, *J. Chromatogr.*, 295, 89, 1984.

[21] G. L. Schmitt and D. J. Pietrzyk, Liquid chromatographic separation of inorganic anions on an alumina column, *Anal. Chem.*, 57, 2247, 1985.

[22] R. W. Pepper, Chemistry Research, 1952, p. 77, Her Majesty's Stationary Office, London, England, 1953.

[23] K. W. Pepper, Sulphonated cross-linked polystyrene: A monofunctional cation-exchange resin, *J. Appl. Chem.*, 1, 124, 1951.

[24] J. R. Parrish, Superficial ion-exchange chromatography, *Nature*, 204, 402, 1965.

[25] J. S. Fritz and J. N. Story, Selectivity behavior of low-capacity, partially sulfonated macroporous beads, *J. Chromatogr.*, 90, 267, 1974.

[26] J. S. Fritz and J. N. Story, Chromatographic separation of metal ions on low-capacity macroreticular resins, *Anal. Chem.*, 46, 825, 1974.

[27] H. Small, Solvent extraction process for the recovery of uranium and rare earth metals from aqueous solutions, U.S. Patent 3,102,782, 1962.

[28] H. Small, Gel liquid extraction–The extraction and separation of some metal salts using tri-n-butylphosphate gels, *J. Inorg. Nucl. Chem.*, 18, 232, 1961.

[29] T. S. Stevens and H. Small, Surface sulfonated styrene divinyl benzene–Optimization of performance in ion chromatography, *J. Liq. Chromatogr.*, 1, 123, 1978.

[30] J. S. Fritz, D. T. Gjerde and R. M. Becker, Cation chromatography with a conductivity detector, Anal. Chem., 52, 1519, 1980.

[31] S. Papanu, C. Pohl and A. Woodruff, New high speed cation exchange columns for ion chromatography, Paper presented at the Pittsburgh Conference and Exposition on Analytical Chemistry and Applied Spectroscopy, Atlantic City, NJ, 1984.

[32] H. A. Battaerd, Core-dual shell graft copolymers with ion exchange resin shells, U.S. Patent 3,565,833, February 23, 1971.

[33] H. A. Battaerd and R. J. Siudak, Synthesis and ion-exchange properties of surface grafts, *J. Macromol. Sci., Chem.*, A4, 1259, 1970.

[34] J. J. Kirkland, Superficially porous chromatographic packing with sulfonated fluoropolymer coating, and chromatographic packing with chemically bonded organic stationary phases, U.S. Patents 3,577,266, May 4, 1971, and 3,722,181, March 27, 1973.

[35] C. G. Horvath, B. A. Preiss and S. R. Lipsky, Fast liquid chromatography: An investigation of operating parameters and the separation of nucleotides on pellicular ion exchangers, *Anal. Chem.*, 39, 1422, 1967.

[36] C. Horvath and S. R. Lipsky, Column design in high pressure liquid chromatography, *J. Chromatogr. Sci.*, 7, 109, 1969.

[37] D. C. Locke, J. T. Schmermund and B. Banner, Bonded stationary phases for chromatography, *Anal. Chem.*, 44, 90, 1972.

[38] D. H. Saunders, R. A. Barford, P. Magidman, L. T. Olszewski and L. T. Rothbart, Preparation and properties of a sulfobenzylsilica cation exchanger for liquid chromatography, *Anal. Chem.*, 46, 834, 1974.

[39] G. J. Sevenich and J. S. Fritz, Preparation of sulfonated gel resins for use in ion chromatography, *Reactive Polymers*, 4, 195, 1986.

[40] G. J. Sevenich and J. S. Fritz, Metal ion selectivity on sulfonated cation-exchange resins of low capacity, *J. Chromatogr.*, 371, 361, 1986.

[41] D. Jensen, J. Weiss, M. A. Rey and C. A. Pohl, Novel weak-acid cation-exchange column, *J. Chromatogr.*, 640, 65, 1993.

[42] J. Morris and J. S. Fritz, Ion chromatography of metal cations on carboxylic acid resins, *J. Chromatogr.*, 602, 111, 1992.

[43] P. Kolla, J. Köhler and G. Schomburg, "Polymer-Coated Cation-Exchange Stationary Phases on the Basis of Silica," *Chromatographia*, 23, No. 7, 465, 1987.

[44] L. M. Nair, R. Saari-Nordhaus and J. M. Anderson, Jr., Simultaneous separation of alkali and alkaline-earth cations on polybutadiene-maleic acid-coated stationary phase by mineral acid eluents, *J. Chromatogr.*, 640, 41, 1993.

[45] L. M. Nair, R. Saari-Nordhaus and J. M. Anderson, Jr., Ion chromatographic separation of transition metals on a polybutadiene maleic-acid-coated stationary phase, *J. Chromatogr. A*, 671, 43, 1994.

[46] R. Saari-Nordhaus, H. Pham and J. M. Anderson, Jr., A new cation stationary phase for challenging ion chromatography applications. Poster 2292, Pittcon '99, Orlando, FL.

4 Detectors

4.1 Introduction

Several types of detectors used in ion chromatographic analysis along with a brief description of how the detectors operate are presented in this chapter. It is important to understand how detectors operate and how ions are detected. For example, detection of iodide in the presence of a salt (sodium chloride) matrix can be accomplished with conductivity detection. But UV detection would be much better due to the fact that iodide will absorb UV light and chloride will not. Because the detection is selective for iodide, the separation conditions can be optimized for rapid interference-free elution.

In IC, the detector must be able to "pick out or see" sample ions in the presence of the eluent ions. There are several methods that can be employed to make this possible. One is to choose a detector that will response only to the sample ions of interest, but not the eluent ions. Another method is to use indirect detection (sometimes called replacement detection). This is where the eluent has a background signal and the presence of samples ions cause a decrease in eluent ions through a replacement process (see Chapter 7). The detector looks for the absence of eluent ions when the sample ion peak elutes and a decreasing signal is detected.

The most widely used method for ion chromatography detection is to treat or choose the eluent prior to detection to make the eluent ions less detectable and or make the sample ions more detectable. The most common example of this is chemical suppression used in conductometric detection. The suppressor is really a membrane chemical reactor that reacts with the post-column eluent stream and changes the ionic counterion for the eluent and for the sample peaks. In its most common form, sodium ions are removed from the stream and hydronium ions are added in an exchange process. This makes the background signal less conducting and the sample signal more conducting. Another example of treatment prior to detection is post-column reaction with a color-forming reagent. PAR, a color-forming chelator, can be added, post-column, to a separation of metal ions to make them detectable by visible spectrophotometric detection. The eluent ions do not react with the color-forming reagent.

General and Selective Detectors: Detectors can be classified either as general or selective. A general detector will respond to all or most of the ions that pass through the detector cell. A conductometric detector is classified as a general detector because all ions will conduct electricity (although to different degrees). UV–VIS spectrophotometric, atomic emission, atomic absorption, and electrochemical detectors can be considered to be selective detectors because they respond only to certain ions. However, any of these can be made into an general type detector by a post-column reaction of sample ions with an appropriate reagent.

General and selective detectors both have their place in ion chromatography. If only one detector were available, a general detector would be most desirable because it would be of the most general use. However, a selective detector can be extremely effective in picking out a single ion of interest (or a small group of ions) from a high eluent background or a high sample matrix ion.

Another key advantage to selective detection is the ability to achieve lower detection limits. A selective detector may or may not have a higher sensitivity (signal per unit concentration) for a particular ion. However, the lower background signals produced by selective detection will translate into lower detection limits because the signal to noise ratio is improved.

4.2 Conductivity Detectors

Conductivity is the ability of an solution containing a salt to conduct electricity across two electrodes. The ability of the solution to conduct is directly proportional to the salt content and the mobility of the individual anions and cations. As the ionic character of a molecule is increased, the conductivity increases. Small, mobile ions conduct quite readily and to a much greater extent than large bulky ions. For example, hydroxide anion is small and mobile and will conduct much better than proprionate anion which larger and bulkier. The relative conductances of ions is shown in Table 4.1.

Table 4.1. Limiting equivalent ionic conductances in aqueous solution at 25 °C. Units: $ohm^{-1} cm^2 equiv^{-1}$ [1].

Angions	λ^-	Cations	λ^+
OH^-	198	H^+	350
F^-	54	Li^+	39
Cl^-	76	Na^+	50
Br^-	78	K^+	74
I^-	77	NH_4^+	73
NO_3^-	71	Mg^{2+}	53
HCO_3^-	45	Ca^{2+}	60
Formate	55	Sr^{2+}	59
Acetate	41	Ba^{2+}	64
Propionate	36	Zn^{2+}	53
Benzoate	32	Hg^{2+}	53
SCN^-	66	Cu^{2+}	55
SO_4^{2-}	80	Pb^{2+}	71
CO_3^{2-}	72	Co^{2+}	53
$C_2O_4^{2-}$	74	Fe^{3+}	68
CrO_4^{2-}	85	La^{3+}	70
PO_4^{3-}	69	Ce^{3+}	70
$Fe(CN)_6^{3-}$	101	$CH_3HN_3^+$	58
$Fe(CN)_6^{4-}$	111	$N(Et)_4^+$	33

Molecular substances such as solvents (water and methanol) and solutions of non-ionized organic acids do not conduct electricity and are not detected by conductivity. The portion of a weak acid that does ionize results in a contribution to the conductivity signal. The ionic form of the weak acid depends on pH. A weak acid with a pKa larger than about 6 cannot be detected with suppressed conductivity detection because all anions are converted to the acid form by the suppressor. So while it is possible to separate arsenite and arsenate on the anion-exchange column (with a high pH eluent), it is only possible to detect the stronger acid from arsenate. It is not possible to detect arsenite by suppressed conductivity detection because it is a anion of too weak an acid and will not conduct once it has passed through the suppressor. Other anions that cannot be detected by suppressed conductivity detection include borate, silicate and cyanide. Non-suppressed, single-column methods, selective detection, or post-column reaction methods are required for the detection of salts of weak acids.

Because of high sensitivity and reduced background noise, the most common form of conductivity detection is with the use of suppressors. The use of suppressors is described in Chapter 6. There is a number of IC users who employ non-suppressed, direct or indirect conductivity detection where the separation column is connected directly to the conductivity cell (Chapter 7). The advantage of this detection method is simplicity in instrument design and operation and the ability to detect salts of weak acids.

Dasgupta and coworkers have proposed a novel approach to simultaneously practicing both suppressed and non-suppressed ion chromatography [2–4]. Several renditions were proposed (Fig. 4.1). The most effective and practical (Fig. 4.1c) is based on first using a conventional NaOH eluent-suppressed IC system. The effluent from detector 1 is combined with a constant concentration of NaOH. The NaOH is introduced by an in-line electrolytic NaOH generator. Then the flow is directed to a second conductivity detector. The second detection background conductance is typically maintained at a level of 20–30 µS/cm corresponding to about 0.1 mM NaOH. The second detector output is the same as what would be observed in a single-column mode with a low concentration NaOH eluent. Together, the two detector outputs provide detection at the microgram per liter level across the pKa range: from fully dissociated to very weak acids up to pKa 9.5.

The two-detector system also provides qualitative information about peak purity and sample pKa. This information is gained by plotting the two detector outputs against each other. In addition to the usual ratio plots, it is suggested that multidimensional detection in any separation system is best served by plotting the raw detector outputs against each other. This type of implementation of single-column and suppressed IC that provides information beyond the sum total of that obtained with either approach alone.

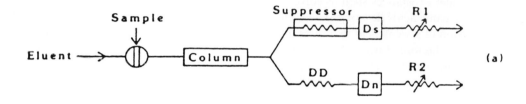

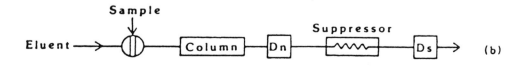

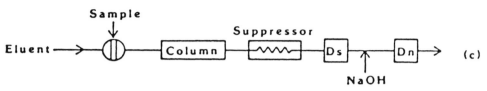

Figure 4.1. Three potential approaches to perform simultaneous, suppressed, and non-suppressed detection. (a) Split stream approach. DD, dummy dispersion device; D_s, suppressed detector; D_n, nonsuppressed detector. Restrictors R1 and R2 are adjusted to provide the same residence time and flow rate in each of the branches. (b) Single column approach. D_n precedes the suppressor. (c) NaOH introduction approach. A small constant quantity of NaOH is introduced after D_s; no restriction is placed on eluent NaOH concentration [4].

4.2.1 Conductivity Definitions and Equations

Electrolytic conductivity is the ability of an electrolytic solution to conduct electricity between two electrodes across which an electric field is applied. Ohm's law, $V = I R$, is obeyed and the magnitude of the current depends, in part, on the magnitude of the applied potential. The conductance, G, of a solution is expressed in terms of the solution electrolytic resistance. It is measured in reciprocal ohms (mhos) or in the SI unit siemens (S).

$$G = 1/R \tag{4.1}$$

Specific conductance (k) takes into account the area of the electrodes (A) in cm^2 and the distance (l) between electrodes, in cm. Conductance increases with the area of the electrodes but decreases as the distance between the electrodes is increased.

$$k = G\,(l/A) \tag{4.2}$$

Thus, k has the units: S cm^{-1}.

The cell constant (K) is equal to l/A in Eq. 4.2, and has the units: cm^{-1}.

$$K = l/A \qquad (4.3)$$

$$k = G\,K \qquad (4.4)$$

Equivalent conductance takes into account the concentration of the chemical solution and is defined by the following equation:

$$\Lambda = \frac{1000\,k}{C} \qquad (4.5)$$

where C is the concentration in equivalents per 1000 cm^3. Λ has the units: S cm^2 equiv^{-1}.

Combining Eqs. 4.4 and 4.5 gives the following equation which relates equivalent conductance to measured conductance, G:

$$G = \frac{\Lambda\,C}{1000\,K} \qquad (4.6)$$

A conductivity detector consists of a detection cell, a readout meter, and the electronics required to measure the conductance and vary the sensitivity setting. The readout for conductance G is given in Siemens (S), or, actually, microSiemens (μS) for solutions that are comparatively dilute.

The specific conductance for a solution can be calculated from the solution conductance (G) if the cell dimensions are known (see Eq. 4.2). However, the usual practice is to measure the conductance of a dilute solution of known specific conductance (such as 0.00100 N KCl) and calculate the cell constant from Eq. 4.4. Once the cell constant is known, the specific conductances of other solutions can be calculated from the measurement of G. The most common cell constant for detectors is 1 or 10; the smaller value is more sensitive.

With a conductivity detector with a known cell constant, the conductances of various solutions of known concentration can be calculated from a table of equivalent conductances, with Eq. 4.6. The limiting equivalent conductances of some common ions are given in Table 4.1. The equivalent conductances of ions generally decrease with increasing concentration because of inter-ionic effects. For dilute solutions (10^{-5} to 10^{-2} N) the equivalent conductances are not greatly different from the values listed in the table.

Example: Calculate the expected conductance of a 1.0×10^{-4} N solution of sodium benzoate in a conductivity cell with a cell constant of 10 cm^{-1}.

From Table 4.1, the equivalent conductance of sodium benzoate will be $\lambda_{Na+} + \lambda_{Bz-}$. Therefore:

$$\Lambda_{Na+Bz-} = 50 + 32 = 82$$

Substituting this into Eq. 4.6:

$$G = (82 \times 1.0 \times 10^{-4})/(1000 \times 10) = 8.2 \times 10^{-7} \text{ mhos or } 0.082\ \mu\text{S}$$

A more typical ion to detect might be chloride. The equivalent conductance is higher so a higher signal would result from this ion, assuming the same peak width and the same sodium counterion. If hydronium ion rather than sodium is the counterion to chloride, then the signal will be multiplied by another factor of 3.4.

4.2.2 Principles of Cell Operation

When an electric field is applied to two electrodes in an electrolytic solution, anions in the solution move toward the anode electrode and cations toward the cathode electrode. The number and the velocities of the ions in the bulk electrolyte determine the resistance of the solution. The ionic mobilities, or the velocity of the ion per unit electric field, depend on the charge and size of the ion, the temperature and type of solution medium, and the ionic concentration.

As the potential that is applied across the electrodes is increased, the ionic velocities increase. Thus, the detector signal is proportional to the applied potential. This potential can be held to a constant value or it can oscillate to a sinusoidal or pulsed (square) wave. Cell current is easily measured; however, the cell conductance (or reciprocal resistance) is determined by knowing the potential to which the ions are reacting. This is not a trivial task. Ionic behavior can cause the effective potential that is applied to a cell to decrease as the potential is applied. Besides electrolytic resistance that is to be measured, Faradaic electrolysis impedance may occur at the cell electrodes resulting in a double layer capacitance. Formation of the double layer capacitance lowers the effective potential applied to the bulk electrolyte.

4.2.3 Conductance Measurement

Techniques involving the use of alternating electrode potentials eliminate the effects of the processes associated with the electrodes. Reversing the polarity of the applied electrode potential reverses the direction of the ion motion, changes the type of electrolysis and changes the type of capacitance formation. The relaxation time (or the ability to recovery) is different for each type of process. As the frequency is increased, effects due to electrolysis are reduced or eliminated and the bulk of current flow is through capacitance formation. An upper frequency limit for detector operation is approximately 1 MHz. At this point the ions cease to move in response to the electric field, although dipole reorientation of the ion electron structure will still occur. Capacitance effects are controlled by matching the cell capacitance in the electronic circuitry or by measuring the instantaneous current. The instantaneous current is the current that is obtained when the potential is first applied and the double layer has not formed.

Some detectors apply a sinusoidal wave potential across the cell electrodes at 100 to 10 000 Hz. A typical detector of this type operates at a frequency of 1 kHz and at a potential of up to 20 V with no electrolysis occurring. The detection method is called synchronous detection. This is when the only current component measured is that current which is in phase with the applied potential frequency. In effect, the measured current flow is always due to an "instantaneous" potential. Other detectors use a bipolar pulse conductance technique [4, 5, 6]. The technique consists of the sequential application of two, short (about 100 μs) voltage pulses to the cell. The pulses are of equal magnitude and duration and opposite polarity. At exactly the end of the second pulse, the cell current is measured and the cell resistance is determined by applying Ohm's law. Because an instantaneous cell current is measured in the bipolar pulse technique, capacitance does not affect the measurement and an accurate cell resistance measurement is made.

Zemann and coworkers have proposed a novel contactless conductivity detector [8, 9]. The detector works without direct contact of the electrode with the eluent or sample. The sensor is based on two metal tubes that are placed around a fused silica capillary with a detection gap of approximately 1.5 mm (Fig. 4.2). A high oscillating frequency of 40–100 kHz is applied to one of the electrodes. A signal is produced on the other electrode as soon as an analyte zone with a different conductivity compared to the background passes through the detection gap. An amplifier and rectifier is connected to the other electrode to measure resistance between the two electrodes. To prevent a capacitance transition between the two electrodes, a thin piece of copper is placed perpendicularly between the electrodes and connected to the ground.

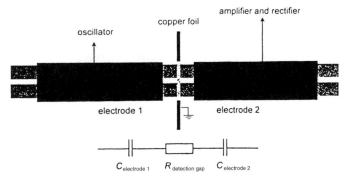

Figure 4.2. Schematic drawing of the contactless capacitively coupled conductivity detector [8].

4.2.4 Hardware and Detector Operation

Conventional conductivity detector cells where the electrolyte is in contact with the electrodes are likely to use electrodes made from 316 stainless steel. A new cell should be treated with 1 N nitric acid for about 60 min to "deactivate" or "passivate" the cell and stabilize the signal. In fact, such nitric acid treatment is a good idea for all parts of a stainless steel IC system.

The mobility of ions in solution varies with the solution temperature. Ionic solutions will increase conductivity about 2 % for every degree increase in temperature. Conductivity detectors usually compensate automatically for temperature change by employing a thermistor monitor and compensation circuitry in which resistance changes linearly with solution temperature. Still, the detector cell (and even the column and tubing) should be placed in an oven for the best detector performance. It is helpful to insulate other components of the ion chromatograph. If the laboratory has large daily temperature swings, temperature control of the column and cell becomes more important. Generally, the instrument temperature is set to at least 5° above the maximum temperature that the laboratory is likely to reach in a given day. Control of the instrument oven and detector temperature and simply keeping the instrument out of drafts is probably the most important single thing that a user can do to contribute to better detector stability.

The flow path may also include a compartment containing the thermistor probe for electronic feedback. Even though conductance cells tend to be low dead volume, the total dead volume in a conductance detector can be quite large and considerable mixing of the eluent stream may take place after the peak has been measured. If two detectors are used, it may be best to place the conductivity detector cell last to avoid peak broadening.

4.3 Ultraviolet–Visible Detectors

There are several reasons why a spectrophotometric detector is a useful detector for monitoring ion-exchange separations. It is selective, yet its selectivity can be changed simply by changing the wavelength monitored by the detector. Versatility of the detector can be increased by adding a color-forming reagent to the eluent or the column effluent. The fundamental law under which ultraviolet–visible (UV–VIS) detectors operate is the Lambert–Beer law. It can be stated in the following form:

$$A = \varepsilon\, b\, C \tag{4.7}$$

A is the absorbance of a species of concentration C, and with an absorptivity ε, in a cell of length b. Concentration is usually in molar concentration and the path length is measured in cm. The term (molar) absorptivity has units that are the inverse of the C and ε units. This leaves A dimensionless; it is usually described in terms of absorbance units. A detector set to a certain sensitivity, for example, 0.16, is said to beat 0.16 Absorbance Units Full Scale sensitivity (0.16 AUFS sensitivity).

The Lambert–Beer equation is useful for choosing conditions for the separation and detection of ions. The eluent ions should have a low absorptivity and the sample ions should have reasonably high absorptivity. In a special case of indirect detection (discussed later), this should be reversed. In this case, the eluent has an absorption signal and the sample is detected by a decrease of the background signal.

It is important to note that when discussing the properties of the eluent and sample ions, it makes a difference whether one is separating anions or cations. For example, if a separation of anions is being discussed, then the absorptivities of the eluent anion and sample anions are considered. But if a low background signal is needed, then of course the cation that is counter to the eluent anions must have a low absorptivity as well.

Some absorbance data are given in this chapter and in some of the applications described in this book. If an ion absorptivity for a particular wavelength is unknown, it can be measured with a spectrophotometer and ion solution of known concentration.

The discussion of UV–VIS detectors for use in ion chromatography is divided into two parts: (a) the direct monitoring of column effluents and (b) post-column derivatization with subsequent spectrophotometric measurement.

4.3.1 Direct Spectrophotometric Measurement

Alkali metal ions are not detected by UV. However, many anions do absorb at lower wavelengths. A list of anions and their detection wavelengths are shown in Table 4.2. The references listed show the different applications of the detection method. The 190 to 210 nm range can be used for the detection of azide, chloride, bromide, bromate, iodide, iodate, nitrite, nitrate, sulfite, sulfide, and selenite. Other work has shown that the 210 to 220 nm range is useful for detecting trithionate, tetrathionate, and pentathionate down to the low nanogram levels. UV detection is particularly useful for anions such as nitrate and iodide which absorb at the longer wavelengths. There have been a number of methods reported for the UV detection of nitrite and nitrate in drinking water and cured meat [37].

Table 4.2. Solutes for direct spectrophotometric detection in IC after ion-exchange separation

Solutes	Wavelength (nm)	Ref.
AsO_4^{3-}, AsO_3^{3-}	200	[10]
$Au(CN)_2^-$	214	[11]
Br^-	195–214	[12]
BrO_3^-	195–210	[10, 13, 14]
$C_2O_4^{2-}$	205	[15]
Citrate	205	[15]
Cl^-	190	[16, 17]
ClO_2^-	195	[10]
ClO_3^-	195	[10]
CN^-	200	[18]
CNO^-	200	[18]
CrO_4^{2-}	365	[19]
Cr^{III}, Cr^{VI}–EDTA	350, 220	[20]
Cr, Pt, Au–Cl^- complexes	225	[21]
$HCOO^-$, CH_3COO^-	190	[16]
I^-	210–235	[22, 23]
IO_3^-	195–210	[10, 14]
Metal Cl^- complexes	210–225	[24–27]
Metal CN^- complexes	210–214	[28, 29]
Metal EDTA complexes	210	[30]

Table 4.2. Continued

Solutes	Wavelength (nm)	Ref.
MoO_4^{2-}	200	[18]
N_3^-	195	[10]
NO_2^-, NO_3^-	200–214	[17, 31, 32]
Organoarsenic acids	210	[33]
PO_4^{3-}	190	[16]
S^{2-}	215	–
$S_2O_3^{2-}, S_3O_6^{2-}, S_4O_6^{2-}$	205	–
SCN^-	195–205	[34, 35]
$SeCN^-$	195	[10]
SeO_3^{2-}	195	[10]
SeO_4^{2-}	195	[10]
SO_3^{2-}	200	[35]
SO_4^{2-}	190	[16, 17]

Aromatic acids absorb well and methods for detecting these anions are powerful.

The usefulness of direct UV detection can be considered to be limited because sulfate is not detected by UV, and chloride, phosphate and others are difficult to detect. Sulfate is probably the most widely analyzed anion by ion chromatography so this is a serious limitation. On the other hand, anions that are difficult to detect make ideal eluent anions.

The absorbance of metal chloride complexes in the ultraviolet spectral region has been used extensively to automatically detect metal ions in liquid chromatography [24–27]. The absorption wavelength maxima of the metal chloride complexes are shown in Table 4.3. Metal EDTA complexes also absorb quite well.

Table 4.3. Complex formation of inorganic anions with ferric perchlorate color-forming reagent[a] [36].

Anion	λ^{max} (nm)	Anion	λ^{max} (nm)
CrO_4^{2-}	305, 344	SO_3^{2-}	308
SCN^-	310	PO_3^{3-}	<300
$Fe(CN)_6^{4-}$	305	$H_2PO_2^-$	<300
$Fe(CN)_6^{3-}$	305	IO_3^-	<300
SO_4^{2-}	306	CO_3^{2-}	<300
Cl^-	335	Br^-	<300
$P_2O_7^{4-}$	310	$B_4O_7^{2-}$	<300
I^-	306, 350	BrO_3^-	<300
$P_3O_{10}^{5-}$	310	CN^-	<300
S^{2-}	<300	SiO_3^{2-}	<300
$S_2O_3^{2-}$	308	NO_3^-	[b]
NO_2^-	372, 360	F^-	[b]
PO_4^{3-}	310	ClO_3^-	[b]

[a] 0.8 M $HClO_4$, 0.05 M $Fe(ClO_4)_3$.
[b] Does not react.

Another technique is indirect detection. In this method, the eluent absorbs strongly in the visible or ultraviolet spectral region. A wavelength is selected where the (usually aromatic) eluent absorbs but the sample ions do not absorb [38–40]. Briefly, because an ion-exchange process is involved, a sample ion can only be eluted by displacement of the eluent ion. This results in a decrease in the signal when a sample ion peak is eluted. While several authors and users have been successful with indirect UV detection, it can be difficult to get the right conditions for separation and detection. Indirect UV detection is generally used only in cases where the separation and detection conditions have been carefully worked out and where a high quality UV detector is available. Temperature control of the column is recommended to control the baseline noise.

4.3.2 Post-Column Derivatization

The post-column method of derivatization of column fractions has been well established from older ion-exchange separations. An appropriate reaction is performed on each fraction to determine the metal ion concentration in that fraction. The automatic addition of a color-forming reagent to an ion-exchange column effluent and analysis by flow-through cell detection is more recent. However, many of the color-forming reagents and buffers used in ion chromatography are the same as those used in the classical fraction method determinations.

The ideal color-forming reagent reacts with a large number of metal ions and has low background absorption. Sickafoose [41] studied the reagents alizarin red S, arsenazo III, chlorophosphonazo III, chrome azurol S, quinalizarin, 4-(2-pyridylazo)-resorcinol (PAR), 4-(2-thiazolylazo)resorcinol (TAR), and xylenol orange. Chrome azurol S and quinalizarin are of limited value, but the other reagents each react with 20 or more metals. PAR is the most general, reacting with 34 metals. Table 4.4 shows a list compiled by Fritz and Story [43] of metals and their reaction with color-forming agents. The 0.0125 % PAR solution was made in 5 M ammonium hydroxide, the 0.00375 % arsenazo III solution was in 2 M ammonium hydroxide and 1 M ammonium acetate, and the arsenazo I solution was in 3 M ammonium hydroxide. Other work [44,45] was done with more dilute PAR solutions (4×10^{-4} M PAR with 3 M ammonium hydroxide and 1 M acetic acid). For lower detection limits (because of lower background signal), the PAR concentration can be reduced even more. But care should be taken not to "overload" the reagent with too high concentrations of sample ion.

Imanari et al. has reported a spectrophotometric detection of many inorganic anions using a post-column reactor [42]. A stream of ferric perchlorate, which is essentially colorless, is mixed with the column effluent. The ferric perchlorate is colorless because perchlorate is a poor complexing anion, but most anions will complex the iron and form colored species that can be detected at 330–340 nm (Table 4.3). A similar detection method works for ions such as orthophosphate, pyrophosphate, nitrilotriacetic acid (NTA), and ethylenediaminetetraacetic acid (EDTA) [46].

Table 4.4. Reactions of color-forming reagents and metals. X denotes a positive reaction [41].

Metal ion	Reagent		
	Arsenazo I	Arsenazo III	PAR
Thorium(lV)		x	x
Zirconium(lV)		x	x
Hafnium(l/)		x	x
Aluminum(III)		x	
Chromium(III)		x	
Lanthanides(III)		x	x
Bismuth(III)			x
Iron(III)			x
Iron(ll)			x
Vanadium(lV)			x
Maganese(ll)			x
Cobalt(ll)			x
Nickel(ll)			x
Copper(ll)		x	x
Zinc(ll)		x	x
Cadmium(ll)			x
Mercury(ll)			x
Lead(ll)		x	x
Magnesium(ll)	x		
Calcium(ll)	x	x	
Barium(ll)		x	
Strontium(ll)		x	

4.3.3 Hardware and Detector Operation

Detectors should have variable wavelength capability. The 190 to 370 nm range is used most often for direct detection. The visible wavelength region is used quite often with the post-column reagents. Metal separations with PAR detection are operated at 520–535 nm and arsenazo III detection at 653 nm. Many of the eluents and other solutions can be corrosive. With highly corrosive solvents, glass or plastic tubing and connectors are required. Several companies now produce commercial post-column reactors specifically designed for IC separations. In addition to this, reactors designed for flow injection analysis (FIA) instrumentation often work. The most important features of any commercial post-column reactor is uniform mixing of the column effluent and the color-forming reagent, and the delivery of the mixture to the detector without broadening the peak. In order to accomplish this, the mixing chamber must be small and efficient. The mixing must be thorough and adapt to any fluctuations in flow from either stream. Sometimes this is not so easy, considering that delivery of the column effluent cannot be easily controlled.

The color-forming reagent concentration should be in large enough excess to keep favorable reaction kinetics and a linear calibration curve. The buffer should be at a high concentration so that reactions are reproducible, rapid, and complete. High buffer concentrations are sometimes necessary to control the pH of a variety of eluents at high concentrations. Because of the high concentrations, reagents may precipitate, tubing and valves may plug, and cell walls may cloud. All solutions should be filtered before use. The instrument may have to be cleaned daily.

4.4 Electrochemical Detectors

Electrochemical detectors are sometimes divided into the groups of potentiometric, amperometric, and conductometric detectors, that is, according to the three parameters of electric measurements. Potentiometric detectors measure voltage, amperometric measure current, and conductometric measure resistance.

Conductometric detectors respond to all ions, but the other detectors respond only to certain electroactive ions. In this book, electrochemical detection will refer to amperometric, or potentiometric detectors, but will not refer to conductometric detectors.

The potentiometric detector operates on the same principles as ion-selective electrodes. An indicating electrode measures a change in the potential in the presence of certain sample ions. Schmuckler et al. [47] explored the use of potentiometric-type detectors for ion chromatography. Halides and pseudohalides were detected with a silver/silver salt indicator electrode. Other indicating electrodes were also suggested, for example, lead/lead salicylate to detect "sulfate type" anions. Other workers have reported potentiometric detection [48–52].

Amperometric detectors are some of the most selective and sensitive detectors used to monitor ion chromatography separations. They are selective because they operate on the principles of oxidation or reduction of substances at an electrode. The potential needed to induce electrolysis differs for each ion. Detector selectivity is controlled by controlling the magnitude of the potential applied to the cell, the detector electrode material and the solution pH. The detectors may operate down to the picoequivalent sample concentration level which requires nanoampere current measurement

Other advantages of electrochemical detection include a wide range of detector response (four to five orders of magnitude) and small cell dead volumes (as low as 1 μL). The detectors are simple, inexpensive, and rugged. Their main disadvantage is that they are sometimes difficult to use. Detectors are sensitive to the eluent flow rate and pH. Trace substances such as oxygen may react depending on the type of electrodes and applied potential. Gold working electrodes are less sensitive to dissolved oxygen than platinum working electrodes. Electrodes can become "poisoned", whereupon their behavior will change.

4.4.1 Detector Principles

Electrochemical detection depends on a catalytic oxidation or reduction at the surface of an electrode located in the column effluent stream. This reaction produces a net charge transfer at the electrode surface. The resulting current is measured and provides a basis for quantification of sample species.

Ions that can be analyzed by electrochemical detection include cyanide, sulfide, hypochlorite, ascorbate, hydrazine, arsenite, phenols, aromatic amines, bromide, iodide, and thiosulfate [53], nitrite and nitrate [54, 55], cobalt and iron [46], and others. The list may be extended through the technique of post-column derivatization to include many more ions such as carboxylic acids, halide ions, alkaline earth ions, and some transition metal ions [57,58]. An example of an electrochemical reaction to detect ions is shown by Eq. 4.8.

$$Ag + X^- \rightarrow AgX + e^- \tag{4.8}$$

where X is a halide sample ion and Ag is the working electrode. A silver electrode can be used to detect fluoride, chloride, bromide, iodide, and cyanide [59].

The reaction rate depends not only on the sample concentration, but also on the applied potential, and on the physical and chemical properties of both the electrode material and the chromatographic eluent. In principle, any chemical species can be detected electrochemically with a suitable electrode potential. In practice, however, the usable potential range is limited to approximately –1.5 V to +1.5 V (see Table 4.5). On the negative end, potentials are limited by the reduction of dissolved oxygen in the eluent and the sample. If the oxygen has been removed from the stream, then the potential is still limited by the reduction of water to hydrogen gas. The limit does vary with the nature of the electrode surface. Table 4.5 compares the working potential range for a number of common electrode types. On the positive end, the working potential range is limited by the hydrolysis of water to liberate oxygen gas. Again, the working limit will vary depending on the surface characteristics of the working electrode used.

Table 4.5. Working potential range[a] of different working electrode materials [59].

Working electrode	Range	
	Basic solution[b] (volts)	Acidic solution[c] (volts)
Glassy carbon[d]	–1.5 to 0.6	–0.8 to 1.3
Gold[e]	–1.25 to 0.75	–0.35 to 1.1
Silver[e]	–1.2 to 0.1	–0.55 to 0.4
Platinum	–0.9 to 0.65	–0.2 to 1.3
Mercury[e]	–1.9 to 0.05	–1.1 to 0.6

[a] With respect to Ag/AgCl reference electrode.
[b] 0.1 N KOH.
[c] 0.1 N HClO$_4$.
[d] The potential limits for glassy carbon do not cut off sharply and must be determined experimentally.
[e] Complexing anions such as halides or cyanide will limit the positive range by several tenths of a volt.

Within this working range, the presence of a reactive sample will give rise to a current/potential (*I/E*) curve (Fig. 4.3). This curve is unique for a particular sample/eluent/electrode combination. It is characterized by a half-wave potential and by a "diffusion current" plateau. The half-wave potential (the potential half way up to the diffusion current plateau) is defined as the potential needed to induce electrolysis of the electroactive species. As the potential is increased, the electrolysis current also increases because more ions migrate to the electrode and become oxidized (or reduced). The electrolysis current eventually forms a plateau in the *I/E* curve because, ultimately, the amount of current is limited by the rate of diffusion of ions to the electrode surface. In normal operation, the electrochemical detector potential is set at the smallest potential possible that is still on the diffusion current plateau. The detector should be on the plateau for consistent performance, but at the lowest potential to lessen the chance of side reactions.

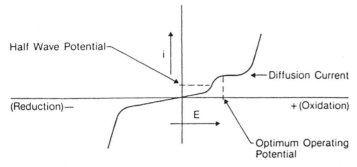

Figure 4.3. The potential needed to induce electrolysis is shown by this current/potential curve. The sample electrolysis results in a diffusion current. This current is proportional to the sample concentration.

Because of the difficulties involved in removing oxygen (and keeping it removed) from the sample and eluent solutions, most applications of electrochemical detectors in ion chromatography have been the detection of relatively easily oxidized species (cyanide sulfide, sulfite, nitrite, or phenols, for example).

Descriptions of detector behavior are often complicated by phenomena such as slow electrolysis kinetics, irreversible reactions, and limits on mass transfer of reactants and products. But ultimately, electrochemical cell behavior can be described by Faraday's law:

$$Q = n F N \tag{4.9}$$

The number of coulombs, Q, measured is related to the number of electroactive moles, N, converted to a product by n, the number of electrons per reaction. F is Faraday's constant. In a detector, the number of coulombs is measured per second, that is, current is measured.

4.4.2 Pulsed Techniques

The activity loss of noble-metal electrodes caused by absorption of organic compounds has plagued electroanalytical chemists since the inception of solid electrode voltammetry [60]. As a response to this problem, Johnson and his students at Iowa State University invented a pulsed-amperometric detector (PAD) that makes constructive use of the adsorption of organic compounds on a noble-metal electrode. Organic molecules are detected by an oxidative desorption process. According to Johnson, virtually any organic that is adsorbed by a cell electrode can be detected with pulsed amperometric detection. Following the detection process, a second potential is used for oxidative cleaning of the electrode surface, and a third potential pulse is then applied to reduce the surface oxide on the electrode. These three steps are illustrated in Fig. 4.4. Sampling of the Faradaic current for the sample detection occurs during the last 50 ms of the E pulse. The time periods at each potential are typically $t_1 = t_2 = 125$ ms and $t_3 = 250$ ms. Hence, the frequency of the waveform is approximately 2 s^{-1}, giving two measurements of the analyte every second. This accurately describes the shape of all but the sharpest of elution peaks. Dionex Corporation has commercialized a PAD detection system for carbohydrate analysis.

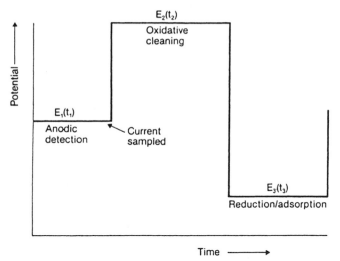

Figure 4.4. Typical waveform for detection of carbohydrates (Courtesy of D.C. Johnson).

From studies on numerous compounds in aqueous alkaline solutions, Johnson concluded that all aldehydes, alcohols, polyalcohols, and carbohydrates are electrochemically detected by the pulsed amperometric scheme described above [61–63]; these compounds include the oligosaccharides [64, 65]. Detection limits are typically less than 0.1 ppm for early eluting peaks when the sample volume is approximately 50 µL.

Johnson has also demonstrated the use PAD will find use for detection of various aromatic compounds as well various aliphatic and aromatic sulfur compounds such as sulfides, disulfides, mercaptans, thiocarbamates, and thiophenes.

There are other instances when a pulsed technique is useful. If species A reacts at a potential $E1$, and both species A and B react at a higher potential $E2$, then a constant potential detector cannot be selective for species B. However, if the detector is pulsed between $E1$ and $E2$, the difference between the current at $E1$ and $E2$ is proportional to the concentration of species B:

$$I_{E2} - I_{E1} \propto [B]$$

Swartzfager reports examples of this procedure [66].

4.4.3 Post-Column Derivatization

Sometimes called secondary electrochemical detection, this technique is useful when the sample species of interest is not electroactive. A reagent is mixed with the column effluent to transform the sample, M, into an electroactive species. An example is shown by Eq. 4.11:

$$[Hg - DPTA]^{3-} + M^{n+} + 2\, e^- \rightarrow [M - DPTA]^{(5-n)-} + Hg \qquad (4.11)$$

The DTPA (diethylenetriaminepentoacetate) metal complex is added and the sample metal "displaces" the mercury ion. The measured current is from the reduction of mercury. There are many other schemes that can be applied to a variety of ions [57, 58].

4.4.4 Hardware and Detector Operation

Electrochemical detectors are constructed with three electrodes. The electrolysis of interest takes place at the working (marked W in Fig. 4.5) electrode at a potential measured by the reference electrode. The auxiliary (or counter) electrode potential is controlled to maintain the reference potential. The reference electrode is usually saturated calomel or silver/silver chloride. Platinum or glassy carbon is generally used for the auxiliary electrode [67, 53].

Choice of the working electrode depends on the sample ions to be determined (Table 4.5). In general, silver is used for halides and pseudohalides. Glassy carbon is useful for oxides and organics. Platinum is probably the most general electrode. It is useful for the oxidation detection of inorganics and organics. Mercury, with its large negative over-potential, is useful for species that can be reduced. Gold is the least susceptible to interference from dissolved oxygen.

Cells are classified according to how the working electrode is positioned relative to the flow stream. There are three major configurations: tubular, thin layer, and wall jet. The tubular cell (open or packed) with its greater working electrode surface area is used for coulometric detection. The thin layer and wall jet designs are used for amperometric detector cells. In thin layer cells, the eluent flow is in the same plane as

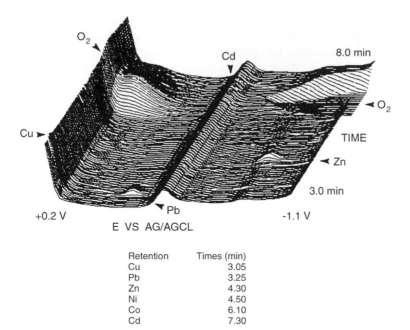

Retention   Times (min)
Cu   3.05
Pb   3.25
Zn   4.30
Ni   4.50
Co   6.10
Cd   7.30

Figure 4.5. Square-wave voltammetric detection of 100 µL of a micromolar mixture of copper, lead, nickel, cobalt, and cadmium. The square wave was 30 Hz with a pulse size of –70 millivolts and a step size of –25 millivolts. Sweep repetition was 2.3 seconds with each sweep on a fresh mercury drop electrode.

the working electrode surface. The wall jet cells are made exactly the same way except that the working electrode is positioned opposite the eluent inlet port. Thus, the eluent "jet" flow is perpendicular to the wall of the working electrode surface. Generally, the wall-jet cell is more sensitive and self-cleaning compared to thin layer cells, but, it is also more flow-sensitive.

Detector response depends on the eluent flow rate, pH, ionic strength, and temperature. In practice, the operating potential of the instrument should be set at the smallest possible value on the diffusion current plateau. Higher potentials do not increase detector response but do increase background noise resulting from the presence of trace contaminants in the eluent. Lower potentials sacrifice some response, but may improve selectivity. Interference may be a problem. Trace substances in the eluent or samples may change detector response. It should be remembered that oxygen can diffuse into the tubing and be traced.

4.5 Refractive Index Detection

Extensive work has been performed with refractive index (RI) detection for ion chromatography [68–70]. The refractive index detector can be considered to be a universal detector because any salt (or acid or base) added to water will cause a change in refractive index of the solvent. The differences in refractive index can be measured as the sample ions pass through the detector window replacing some of the eluent ions and changing the refractive index. Depending on the relative change, the peak may be increasing or decreasing. The detector has been used for both anion and cation separations with good results.

The refraction index of solution increases with the molecular weight (and concentration) of the solute dissolved in the solution. Organic ions usually cause a greater change than inorganic ions in the refractive index of a solution. Refractometers can be used for indirect detection if the eluent ion is organic, for example, phthalate. Sample peaks of most sample ions would be seen as decreasing refractive index peaks. Direct Rl detection may be possible when the sample ions are organic. The eluent in this case would probably be inorganic.

Refractive index detection allows an extremely wide latitude in the selection of the eluent type, eluent pH and the ionic strength. In principle, refractive index detection can be substituted for conductance or UV absorption detection in many separations. However, in early work, refractive detection was found to be only moderately sensitive and was considered to be somewhat interference-prone [71]. Minimum detectable quantities for common anion such as chloride nitrate, or sulfate were reported to be in the 20 ng to 50 ng range (compared with 1 to 5 ng for direct conductance detection).

The stability of Rl detection has improved dramatically and it should be considered seriously as an option for rugged, sensitive detection. This is due, at least partly, to control of the cell temperature, improved electronics and improved optical transducers. The new detectors are much less sensitive to variations in room temperature.

The highest sensitivity levels can be reached only with careful control of the chromatographic temperature. Therefore column ovens should be considered when using RI.

4.6 Other Detectors

Many other detectors have been used to monitor ion chromatography separations. Most of these detectors have been used only in special cases. Flame photometric detection [72, 73] has been used to detect alkali, alkaline earth, and some rare earth metals. Atomic absorption (AA) detectors [74–76] have been used for arsenite, arsenate, monomethyl arsenate, dimethyl arsinate, and *p*-aminophenoarsenate separations. Detectors of this type can be extremely sensitive detecting arsenic down to 10 ng/mL.

Research developing environmental methods might consider the use of AA detection. Perhaps some of the more interesting are detectors that use inductively coupled plasma (ICP) as an energy source and either atomic emission (AE) or mass spectrometry (MS) as the detector. ICP-AE and ICP-MS are well-developed analytical tools. Once of the major advantages of these techniques is that mixture of metals can be analyzed without the need for separation. Thus, workers who use these instruments normally do not think about their use as detectors. However, ICP-AE and ICP-MS cannot determine the oxidation or chemical state of a particular metal ion. Some samples are quite important from a toxicological and environmental standpoint since the toxicity of a metal may depend on its oxidation state. For example, Cr (III) is not toxic (and even considered an essential element), but Cr(VI) is extremely toxic. Inductively coupled plasma atomic emission [77] detector was used to detect rare earth metals.

References

[1]   R. N. Reeve, Determination of inorganic main group anions by high-performance liquid chromatography, *J. Chromatogr.*, 177, 393, 1979.
[2]   I. Berglund and P. K. Dasgupta, Two-dimensional conductometric detection in ion chromatography: post suppressor conversion of elute acids to a salt, *Anal. Chem.*, 64, 3007, 1991.
[3]   I. Berglund and P. K. Dasgupta, Two-dimensional conductometric detection in ion chromatography: postsuppressor conversion of elute acids to a base, *Anal. Chem.* 63, 2175, 1991.
[4]   I. Berglund, P. K. Dasgupta, J. Lopez, and O. Nara, Two-dimensional conductometric detection in ion chromatography: sequential suppressed and single column detection, *Anal. Chem.*, 65, 1192, 1993.
[5]   D. E. Johnson and C. G. Enke, Bipolar pulse technique for fast conductance measurements, *Anal. Chem.*, 42, 329, 1970.
[6]   K. J. Caserta, F. J. Holler, S. R. Crouch, and C. G. Enke, Computer controlled bipolar pulse conductivity system for applications in chemical rate determinations, *Anal. Chem.*, 50, 1534, 1978.
[7]   W. A. McKinley, Tracor Inst., Austin Texas, Pittsburgh Conference, Paper No. 309 (1981).
[8]   K. Mayrhofer, A. J. Zemann, E. Schnell, and G. K. Bonn, Capillary electrophoresis and contactless conductivity detection of ions in narrow inner diameter capillaries, *Anal. Chem.*, 71, 3828, 1999.
[9]   A. J. Zemann, E. Schnell, D. Volgger, and G. K. Bonn, Contactless conductivity detection for capillary electrophoresis, *Anal. Chem.* 70, 563, 1998.
[10]   R. J. Williams, Determination of inorganic anions by ion chromatography with ultraviolet absorbance detection, *Anal. Chem.*, 55, 851, 1983.
[11]   P. R. Haddad and N. E. Rochester, Ion-interaction reversed-phase chromatographic method for the determination of gold (I) cyanide in mine process liquors using automated sample preconcentration, *J. Chromatogr.*, 439, 23, 1988.
[12]   Waters Ion Brief No. 88112.
[13]   P. R. Haddad and P. E. Jackson, The determination of ascorbate, bromate and metabisulfite in bread improvers using high performance ion-exchange chromatography, *Food Tech. Aust.*, 37, 305, 1985.
[14]   Wescan Application #164.
[15]   D. R. Jenke, Quantitation of oxalate and citrate by ion chromatography with a buffered, strong acid eluent, *J. Chromatogr.*, 437, 231, 1988.
[16]   G. P. Ayers and R. W. Gillett, Sensitive detection of anions in ion chromatography using UV detection at wavelengths less than 200 nm, *J. Chromatogr.*, 284, 510, 1984.
[17]   L. Eek and N. Ferrer, Sensitive determination of nitrite and nitrate by ion-exchange chromatography, *J. Chromatogr.*, 322, 491, 1985.
[18]   T. E. Boothe, A. M. Emran, R. D. Hnn, P.J. Kothari and M.M.Vora, Chromatography of radiolabeled anions using reversed-phase liquid chromatographic columns, *J. Chromatogr.*, 333, 269, 1985.
[19]   Dionex Application Note 51.
[20]   A.F. Geddes and J.G. Tarter, The ion chromatographic determination of chromium(III)-chromium(VI) using an EDTA eluant, *Anal. Lett.*, 21, 857, 1988.
[21]   D.T. Gjerde and J.S. Fritz, Chromatographic separation of metal ions on macroreticular anion-exchange resins of a low capacity, *J. Chromatogr.*, 188, 391, 1980.
[22]   R.L. Smith, Z Iskandarani and D.J. Pietrzyk, Comparison of reversed stationary phases for the chromatographic separation of inorganic analytes using hydrophobic ion mobile phase additives, *J. Liq. Chromatogr.*, 7, 1935, 1984.
[23]   I. Molnar, H. Knauer and D. Wilk, High-performance liquid chromatography of ions, *J. Chromatogr.*, 201, 225, 1980.
[24]   M. D. Seymour, J. P. Sickafoose, and J. S. Fritz, Application of forced-flow liquid chromatography to the determination of iron, *Anal. Chem.*, 43, 1734, 1971.
[25]   M. D. Seymour and J. S. Fritz, Rapid, selective method for lead by forced-flow liquid chromatography, *Anal. Chem.*, 45, 1632, 1973.

[26] J. S. Fritz and L. Goodkin, Separation and determination of tin by liquid-solid chromatography, *Anal. Chem.*, 46, 959, 1974.

[27] L. Goodkin, M. D. Seymour, and J. S. Fritz, Ultraviolet spectra of metal ions in 6M hydrochloric acid, *Talanta*, 22, 245, 1975.

[28] B. Grigorova, S.A. Wright and M. Josephson, Separation and determination of stable metallo-cyanide complexes in metallurgical plant solutions and effluents by reversed-phase ion-pair chromatography, *J. Chromatogr.*, 410, 19, 1987.

[29] D.F. Hilton and P.R. Haddad, Determination of metal-cyano complexes by reversed-phase ion-interaction high-performance liquid chromatography and its application to the analysis of precious metals in gold processing solutions, *J. Chromatogr.*, 361, 141, 1986.

[30] S. Matsushita, Simultaneous determination of anions and metal cations by single-column ion chromatography with ethylenediaminetetraacetate as eluent and conductivity and ultraviolet detection, *J. Chromatogr.*, 312, 327, 1984.

[31] J. Osterloh and D. Goldfield, Determination of nitrate and nitrite ions in human plasma by ion-exchange high-performance liquid chromatography, *J. Liq. Chromatogr.*, 7, 753, 1984.

[32] Waters Ion Brief No. 88111.

[33] M. Marno, N. Hirayama, H. Wada and T. Kuwamoto, Separation and determination of organoarsenic compounds with a microbore column and ultraviolet detection, *J. Chromatogr.*, 466, 379, 1989.

[34] Y. Michigami, T. Takahashi, F. He, Y. Yamamoto and K. Ueda, Determination of thiocyanate in human serum by ion chromatography, *Analyst* (London), 113, 389, 1988.

[35] A. Mangia and M.T. Lugari, Separation and determination of inorganic anions by means of ion-pair chromatography, *Anal. Chim. Acta*, 159, 349, 1984.

[36] R. L. Cunico and T. Schalbach, Comparison of ninhydrin and o-phthalaldehyde post-column detection techniques for high-performance liquid chromatography of free amino acids, *J. Chromatogr.*, 266, 461, 1983.

[37] P.E. Jackson P.R. Haddad and S. Dilli, Determination of nitrate and nitrite in cured meats using high-performance liquid chromatography, *J. Chromatogr.*, 295, 471, 1984.

[38] M. Denkert, L. HacEzell, G. Schill, and E. Sjogren, Reversed-phase ion-pair chromatography with UV-absorbing ions in the mobile phase, *J. Chromatogr.*, 218, 31, 1981.

[39] H. Small and T. E. Miller, Jr., Indirect photometric chromatography, *Anal. Chem.*, 54, 462, 1982.

[40] R. A. Cochrane and D. E Hillman, Analysis of anions by ion chromatography using ultraviolet detection, *J. Chromatogr.*, 241, 392, 1982.

[41] J. P. Sickafoose, Ph. D. Dissertation, Iowa State University Ames, Iowa (1971).

[42] T. Imanari, S. Tanabe, T. Toida, and T. Kawanishi, High-performance liquid chromatography of inorganic anions using iron(3+) as a detection reagent, *J. Chromatogr.*, 250, 55, 1982.

[43] J. S. Fritz and J. N. Story, Chromatographic separation of metal ions on low capacity, macroreticular resins, *Anal. Chem.*, 46, 825, 1974.

[44] S. Elchak and R. M. Cassidy, Separation of the lanthanides on high-efficiency bonded phases and conventional ion-exchange resins, *Anal. Chem.*, 51, 1434, 1979.

[45] C. H. Knight, R. M. Cassidy, B. M. Recoskie, and L. W. Green, Dynamic ion exchange chromatography for determination of number of fissions in thorium-uranium dioxide fuels, *Anal. Chem.*, 56, 474, 1984.

[46] A. W. Fitchett and A. Woodruff, Determination of polyvalent anions by ion chromatography, *L. C.*, 1, 48, 1983.

[47] H. Hershcovitz, Ch. Yarnitsky, and G. Schmuckler, Ion chromatography with potentiometric detection, *J. Chromatogr.*, 252, 113, 1982.

[48] P. R. Haddad, P. W. Alexander, and M. Trojanowicz, Ion chromatography of magnesium, calcium, strontium and barium ions using a metallic copper electrode as a potentiometric detector, *J. Chromatogr.*, 294, 397, 1985.

[49] P. R. Haddad, P. W. Alexander, and M. Trojanowicz, Ion chromatography of inorganic anions with potentiometric detection using a metallic copper electrode, *J. Chromatogr.*, 321, 363, 1985.

[50] P. W. Alexander, P. R. Haddad, and M. Trojanowicz, Potentiometric detection in ion chromatography using a metallic copper indicator electrode, *Chromatographia*, 20, 179, 1985.

[51] P. R. Haddad, P. W. Alexander, and M. Trolanowicz, Application of indirect potentiometric detection with a metallic copper electrode to ion chromatography of transition metal ions, *J. Chromatogr.*, 324, 319, 1985.

[52] K. Suzuki, H. Aruga, H. Ishiwada, T. Oshima, H. Inoue and T. Shirai, Determination of anions with a potentiometric detector for ion chromatography, *Bunseki Kagaku*, 32, 585, 1983 (Japanese).

[53] Dionex Corporation, "Electrochemical Detector", Sunnyvale, CA, February, 1981.

[54] R. J. Davenport and D. C. Johnson, Determination of nitrate and nitrite by forced-flow liquid chromatography with electrochemical detection, *Anal. Chem.*, 46, 1971, 1974.

[55] G. A. Sherwood and D. C. Johnson, A chromatographic determination of nitrate with amperometric detection at a copperized cadmium electrode, *Anal. Chim. Acta*, 129, 101, 1981.

[56] Y. Takata, F. Mizaniwa, and C. Moekoya, Electrocatalytic detection of cobalt separated by ion exchange chromatography, *Anal. Chem.*, 51, 2337, 1979.

[57] Y. Takata and G. Muto, Flow coulometric detector for liquid chromatography, *Anal. Chem.*, 45, 1864, 1973.

[58] J. E. Girard, Ion chromatography with coulometric detection for the determination of inorganic ions, *Anal. Chem.*, 51, 836, 1979.

[59] R. D. Rocklin, Working-electrode materials: proper selection for detectors used in liquid and ion chromatography, *L. C.*, 2, 588, 1984.

[60] R. N. Adams, "Electrochemistry at Solid Electrodes", Marcel Dekker Inc., New York (1969), Chapter 7.

[61] S. Hughes, P. L. Meshi, and D. C. Johnson, Amperometric detection of simple alcohols in aqueous solutions by application of a triple-pulse potential waveform at platinum electrodes, *Anal. Chim. Acta*, 132, 1, 1981.

[62] S. Hughes and D. C. Johnson, Amperometric detection of simple carbohydrates at platinum electrodes in alkaline solutions by application of a triple-pulse potential waveform, *Anal. Chim Acta*, 132, 11, 1981.

[63] S. Hughes and D. C. Johnson, High-performance liquid chromatographic separation with triple-pulse amperometric detection of carbohydrates in beverages, *J. Agric. Food Chem.*, 30, 712, 1982.

[64] R. D. Rocklin and C. A. Pohl, Determination of carbohydrates by anion exchange chromatography with pulsed amperometric detection, *J. Liq. Chromatogr.*, 6, 1577, 1983.

[65] M. Malfoy and J. A. Renaud J. Electroanal., Electrochemical investigations of amino acids at solid electrodes. Part II. Amino acids containing no sulfur atoms: tryptophan, tyrosine, histidine and derivatives, *Chem. Interfacial Electrochem.*, 114, 213, 1980.

[66] D. G. Swartzfager, Amperometric and differential pulse voltammetric detection in high performance liquid chromatography, *Anal. Chem.*, 48, 2189, 1976.

[67] R. J. Rucki, Electrochemical detectors for flowing liquid systems, *Talanta*, 27, 147, 1980.

[68] Chrompack, Inc., Bridgewater, New Jersey, Chrompack Topics, Vol. 8 (1981).

[69] F. A. Buytenhuys, Ion chromatography of inorganic and organic ionic species using refractive index detection, *J. Chromatogr.*, 218, 57, 1981.

[70] P. R. Haddad and A. L. Heckenberg, High-performance liquid chromatography of inorganic and organic ions using low-capacity ion-exchange columns with indirect refractive index detection, *J. Chromatogr.*, 252, 177, 1982.

[71] T. Jupille, UV-visible absorption derivatization in liquid chromatography, *J. Chromatogr. Sci*, 17, 160, 1979.

[72] D. J. Freed, Flame photometric detector for liquid chromatography, *Anal. Chem.*, 47, 186, 1975.

[73] S. W. Downey and G. M. Hieftje, Replacement ion chromatography with flame photometric detection, *Anal. Chim. Acta*, 153, 1, 1983.

[74] E. A. Woolson and N. Aharonson, Separation and detection of arsenical pesticide residues and some of their metabolites by high pressure liquid chromatography-graphite furnace atomic absorption spectrometry, *J. Assoc. Off. Anal. Chem.*, 63, 523, 1980.

[75] G. R. Ricci, L. S. Shepard, G. Colovos, and N. E. Hester, Ion chromatography with atomic absorption spectrometric detection for determination of organic and inorganic arsenic species, *Anal. Chem.*, 53, 610, 1981.

[76] A. A. Grabinski, Determination of arsenic(III), arsenic(V), monomethylarsonate, and dimethylarsinate by ion-exchange chromatography with flameless atomic absorption spectrometric detection, *Anal. Chem.*, 53, 966, 1981.

[77] K. Yoshida and H. Haragachi, Determination of rare earth elements by liquid chromatography/ inductively coupled plasma atomic emission, *Anal. Chem.*, 56, 1984.

5 Principles of Ion Chromatographic Separations

5.1 Basic Chromatographic Considerations

The discussions in this chapter describing the principles of chromatographic performance are illustrated with data and equations describing cation exchange behavior. The same principles also apply to anion exchange behavior.

5.1.1 Chromatographic Terms

Since ion chromatography is really a form of column liquid chromatography, it follows that the same chromatographic terms should be used. The major terms are summarized in Fig. 5.1 and Table 5.1. The picture is unfortunately clouded by the fact that not everyone uses the same name or exactly the same symbol for some of these terms. For example, it is now recommended to use the name *retention factor* for what was called the *capacity factor* for many years. Both k and k' have been used as the symbol

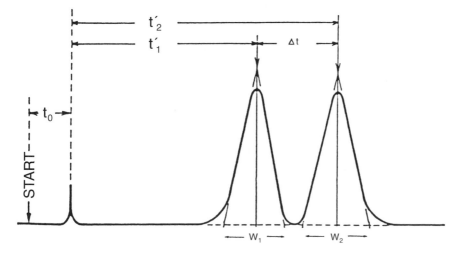

Fig. 5.1. Some common terms used in chromatography.

for this term. A similar problem exists with the time required for a nonsorbed marker to pass through the chromatographic system from the point of injection to the point of detection. Some refer to this quantity as the dead time, t_o, while others call it the time of mobile phase to pass, t_M. Although the retention time (the time for a sample component to be eluted from the column to its peak maximum) is given the symbol t_R, we prefer to simply write it as t. This makes it possible to denote the retention times of several peaks as $t_1, t_2, t_3, ...$

Table 5.1. Chromatographic terms.

Term	Symbol	Definition
Retention time	t or t_R	Time to elute a peak to its maximum concentration
Dead time	t_o or t_M	Time to elute a non-sorbed marker
Adjusted retention time	t'	$t' = t - t_o$
Retention factor (or capacity factor)	k or k'	$k = (t - t_o)/t_o$
Peak width	w	Peak width at its base, $w = 4\sigma$
Peak width at 1/2 peak height	$w_{1/2}$	$w_{1/2} = 2.35\sigma$
Peak resolution	R_s	$R_s = (t_2 - t_1)/w = \Delta t/4\sigma$
Separation factor	α	$\alpha = k_2/k_1 = t'_2/t'_1$
Theoretical plate no.	N	$N = t^2/\sigma^2 = 16t^2/w^2$
Height equivalent of a theoretical plate	H	$H = L/N$ (L = column length)

Most chromatographers continue to use N (or sometimes n) to denote the number of theoretical plates, which is used as a measure of the separation power of a particular chromatographic column and system. The larger N is, the greater the separation power. The height equivalent of a theoretical plate H (or sometimes h) denotes the separation efficiency of a column. Thus, a lower value for H denotes a more efficient column. Although it is well known that H (and therefore N) changes with the linear flow rate (u) of the mobile phase, it should be pointed out that H and N also vary with the retention factors of the analytical sample components. Analytes with low k values generally have higher plate numbers (sharper, better resolved peaks) than those with higher k values (4 to 10, for example).

A better way to look at chromatographic efficiency is to consider the dynamic processes that contribute to peak broadening. Recall that chromatographic peaks are Gaussian and that the width of a peak at its base is approximately four standard deviations: $w = 4\sigma$. The factors that contribute to peak broadening are *additive* provided the variance (σ^2) is used instead of the standard deviation. The total variance (σ^2_{tot}) is the sum of variances due to multipaths (σ^2_{mp}), axial diffusion (σ^2_{dif}), resistance to mass transfer (σ^2_{mt}) and extra-column (σ^2_{ec}).

$$\sigma^2_{tot} = \sigma^2_{mp} + \sigma^2_{dif} + \sigma^2_{mt} + \sigma^2_{ec} \tag{5.1}$$

The differences in tortuous paths through the packed column (i.e., multipaths) is minimized by using spherical solid particles of a very narrow size range, packed very evenly into the column. The value of σ^2_{mp} will increase with the average diameter of the packing material.

The second term, σ^2_{dif}, represents the zone spreading that each sample component exhibits due to diffusion along the column axis. Diffusion coefficients in aqueous solution are generally low, so the contribution of this term is relatively small unless the retention time is quite long due to a very slow flow rate or a high retention factor.

Resistance to mass transfer, σ^2_{mt}, is by far the major contributor to sample zone spreading within the column. This term is minimized by a column packing that attains equilibrium of the analyte between the mobile and stationary phases as quickly as possible. A moderate linear flow rate, u, should be employed because σ^2_{mt} increases with flow rate.

The last term, σ^2_{ec}, points out that substantial peak broadening may occur *outside* the column. To avoid this, the transfer lines from sample injection to column and column should be as short as possible. Stagnant areas in the system must be avoided. These can occur, for example, if the connection between two pieces of tubing bows out. The detector cell should have a low dead volume.

5.1.2 Retention Factors

Perhaps the single most important term in any kind of chromatography is the retention factor (or capacity factor), k. Conditions must be adjusted so that there is a sufficient difference in the k values of the various analytes to give a good separation. It is also necessary to select conditions so that the range of k values is such that a separation may be completed within a reasonable time. A k range of 2 to 10 has often been specified as desirable.

A satisfactory chromatographic separation depends on having a column with a sufficient plate number, N, as well as an adequate difference in k values. Resolution in terms of separation factor ($\alpha = k_2/k_1$), the average retention factor, $k_{av} = (k_1 + k_2)/2$, and the plate number, N, is given by:

$$R = \frac{\alpha-1}{\alpha+1} \cdot \frac{\sqrt{N}}{2} \cdot \frac{k_{av}}{1+k_{av}}$$

(5.2)

This relationship is often used to estimate the number of plates needed for a separation.

$$\sqrt{N} = \frac{2(\alpha+1)}{\alpha-1} \cdot \frac{1+k_{av}}{k_{av}} \cdot R_s$$

(5.3)

For example, if $k_2 = 5.2$ and $k_1 = 4.8$, $\alpha = 1.08$, for $R_s = 1.0$

$$\sqrt{N} = 2 \frac{2.08}{0.08} \cdot \frac{6}{5} \cdot 1.0 = 62.4$$

In reversed-phase liquid chromatography, separations are based on differences in partitioning of analytes between a rather hydrophobic stationary phase and a mobile liquid phase such as acetonitrile–water. Retention factors generally increase as the molecular size and hydrophobicity of the organic analytes become larger. The values of k become smaller as the proportion of organic solvent in the mobile phase is increased. The retention factors of the analytes to be separated are kept in a desirable range by adjusting the composition of the mobile phase.

In ion chromatography the retention factor of an analyte is again determined by its relative affinity for the stationary and mobile phases but the mechanism is different. The stationary phase is an ion exchange material, and the analyte ions are retained only at specific ionic sites on the ion exchanger. In order for the analyte to be attracted to the exchange site, an eluent ion of the same charge (positive or negative) must be displaced. The retention factor of an the analyte ion is kept within the desired range by adjusting the concentration of the competing ion in the eluent.

A specific example may be used to illustrate these principles. A cation exchanger (Catex) is converted to the H^+ form by passing a solution of hydrochloric acid through the ion-exchange column. Introduction of a sample containing Na^+ as an the analyte then sets up the exchange equilibrium:

$$Catex - H^+ + Na^+ \rightleftarrows Catex - Na^+ + H^+ \tag{5.4}$$

By adjusting the H^+ concentration in the eluent to an appropriate value (0.01 M in some instances), the ratio of Catex – Na^+ to Na^+ in solution, and hence the retention factor, can be kept within desired limits.

5.2 Ion-Exchange Equilibria

5.2.1 Selectivity Coefficients

Ion-exchange reactions are reversible; therefore, ion-exchange behavior can be described in terms of equilibrium equations. The exchange of two ions, A and B, competing for reaction with the resin is represented by Eq. 5.5a for monovalent ions and by Eq. 5.5b for ions of different charge.

$$A_s + B_r \rightleftarrows A_r + B_s \tag{5.5a}$$

$$bA_s + aB_r \rightleftarrows bA_r + aB_s \tag{5.5b}$$

In these equations, the subscript s denotes the solution phase and r the resin phase. The equilibrium constant for each of these equilibria is as follows:

$$K_B^A = \frac{[A]_r [B]_s}{[B]_r [A]_s} \tag{5.6a}$$

$$K^A_B = \frac{[A]_r{}^b [B]_s{}^a}{[B]_r{}^a [A]_s{}^b}$$ (5.6b)

The brackets indicate the ion concentration in mmol/mL for the solution phase and in mmol/g for the resin phase. For accurate calculations of equilibrium constants, we should really use the activities (to calculate the effective concentration) of the ions instead of their concentrations. However, activity coefficients are sometimes difficult to measure, especially within the ion-exchange resin matrix. In ion chromatography, ionic concentrations are often low and ion activities approach unity. For the sake of simplicity, only ion concentrations are used in this discussion.

The equilibrium constant in Eq. 5.6 is called the *selectivity coefficient*. A large value for K^A_B means that the resin has a higher affinity for the A ion than for the B ion. Tables 5.2 and 5.3 show selectivity coefficients for several cations on AG 50WX8 and for some anions on Dowex 1X8, respectively.

Table 5.2. Selectivity coefficients, K^A_B, for cations on AG 50WX8 cation exchange resin. Here, B represents H^+.

Cation	Selectivity coefficient
H^+	1.0
Li^+	0.9
Na^+	1.5
$NH_4{}^+$	2.0
K^+	2.5
Rb^+	2.6
Cs^+	2.7
Cu^+	5.3
Ag^+	7.6
Mn^{2+}	2.4
Mg^{2+}	2.5
Fe^{2+}	2.6
Zn^{2+}	2.7
Co^{2+}	2.8
Cu^{2+}	2.9
Cd^{2+}	3.0
Ni^{2+}	3.0
Ca^{2+}	3.9
Sr^{2+}	5.0
Hg^{2+}	7.2
Pb^{2+}	7.5
Ba^{2+}	8.7

These data are for classical resins of high exchange capacity. Selectivity coefficients for the resins used in modern ion chromatography are apt to be significantly different. Nevertheless, the values in Tables 5.2 and 5.3 do give some indication of the relative affinity of the resins for various ions.

Selectivity coefficients can be used to estimate the effectiveness of different ions as eluents in ion chromatography. Ions with high selectivity coefficients usually make the most efficient eluents. This means that a relatively dilute solution of the eluting ion can be used. Sometimes an eluent will elute the sample ions too quickly, even after further dilution of the eluent. In some cases it may be necessary to choose an eluent ion that has a lower selectivity coefficient. As a general rule, divalent eluent ions are best for separating divalent sample ions and monovalent eluents are used for separating monovalent sample ions.

Table 5.3. Selectivity coefficients, K^A_B, for anions on Dowex 1x8. Here, B represents Cl$^-$

Anion	Selectivity coefficient
Salicylate	32.2
Iodide	8.7
Phenoxide	5.2
Bisulfate	4.1
Nitrate	3.8
Bromide	2.8
Cyanide	1.6
Bisulfite	1.3
Nitrite	1.2
Chloride	1.00
Bicarbonate	0.32
Dihydrogen Phosphate	0.25
Formate	0.22
Acetate	0.17
Aminoacetate	0.10
Fluoride	0.09
Hydroxide	0.09

5.2.2 Other Ion-Exchange Interactions

In addition to true ion exchange, other interactions can take place between the sample solutes and the resin. Adsorption is one of the commonest of these interactions. For example, the benzoate anion appears to be adsorbed somewhat by the polystyrene–divinylbenzene polymeric matrix of organic ion exchangers. This may be due to an attraction of the π electrons of the aromatic polymer for the benzoate. Benzoic acid, which exists mostly in the molecular form, is absorbed to a much greater degree than benzoate salts.

Adsorption of eluent and sample ions by an ion exchange resin appears to be quite common, although the degree to which adsorption occurs varies tremendously. Large organic ions and some of the larger inorganic ions seem to be appreciably adsorbed, while adsorption of small, polar ions may be negligible.

Much of the selectivity in separating mixtures of organic cations and anions has been shown to come from differences in adsorption rather than from differences in ion exchange selectivity [1,2]. However, adsorption and subsequent desorption is apt to be a slower process than ion exchange and is therefore to be avoided as much as possible in ion chromatography.

5.2.3 Distribution Coefficient

Distribution coefficients, like selectivity coefficients, are a measure of the affinity of the resin for a particular solute ion. A large distribution coefficient indicates that an ion-exchanger has a greater affinity of an ion.

The weight distribution coefficient, D_g, for the exchange ion A, is given by Eq. 5.7:

$$D_g = \frac{[A]_r}{[A]_s} = \frac{(mmol\ A_r)(mL\ solution)}{(mmol\ A_s)(g\ resin)} \tag{5.7}$$

Some workers have used a volume distribution coefficient, D_v.

$$D_v = D_g\ d \tag{5.8}$$

where d is the resin bed density.

$$d = gram\ of\ dry\ resin/mL\ resin\ bed\ volume \tag{5.9}$$

Experimentally, the weight distribution coefficient of an ion A in a given eluent is measured by equilibrating a known volume of standard solution with a known weight of resin and calculating the results from Eq. 5.7. The volume distribution coefficient, D_v, of an ion is calculated from its retention volume from a column with a given eluent:

$$V_R = D_v + 1 \tag{5.10}$$

5.2.4 Retention Factor

In modern chromatography it is more convenient to use the capacity factor (k or k') instead of the distribution coefficient. The name *retention factor* has been suggested to replace the name capacity factor. The definition for capacity factor or retention factor is the same:

$$k = \frac{amount\ of\ analyte\ in\ column\ stationary\ phase}{amount\ of\ analyte\ in\ column\ mobile\ phase} \tag{5.11}$$

The most convenient way to calculate k is from the retention time (t_R) of a sample ion and from the dead time (t_o or t_M), measured from a chromatogram. The value of t_o (also called t_M) is the time for a nonretained substance to pass through the chromatographic column and detector. The substance chosen can be either a detectable ion or molecule, but care must be taken to choose a substance that passes through the column with absolutely no retention.

Once t_R and t_o have been measured, the k for that ion can be calculated from the equation:

$$k = (t_R - t_o)/t_o \tag{5.12}$$

Alternatively, the retention volume (V_R) and the dead volume (V_o) can be used.

$$k = (V_R - V_o)/V_o \tag{5.13}$$

The retention factor is a useful constant for any given analyte and chromatographic system, because it does not vary with column length or flow rate. For example, as column length is increased, t_o and t_R will both increase, but k' does not change. The retention time for any analyte follows the simple relationship:

$$t_R = t_o(1 + k) \tag{5.14}$$

The adjusted retention time, $t'_R = t_R - t_o$, is also frequently used in chromatography.

Suppose we have an ion-exchange column that has been equilibrated with an eluent ion with a positive or negative charge x (E^x). The exchange reaction with a sample ion of charge y (I^y) will be as follows:

$$x\,I_{soln} + y\,E_{res} \rightleftarrows x\,I_{res} + y\,E_{soln} \tag{5.15}$$

The equilibrium constant for this exchange is:

$$K = [I_{res}]^x\,[E_{soln}]^y/[I_{soln}]^x\,[E_{res}]^y \tag{5.16}$$

where the brackets denote activities.

The capacity factor, k, will be proportional to the ratio of I on the resin phase to that in solution. The amount of E in the resin will remain virtually constant because less than 1 % of the total ion exchange capacity is typically used in ion chromatography. This term will be replaced by the resin ion-exchange capacity, C. Making these substitutions:

$$K = k^x E^y/C^y \tag{5.17}$$

Solving for k, converting to logs, and setting K equal to a constant for any given ion:

$$x \log k = -y \log [E] + y \log C + \text{Constant} \tag{5.18}$$

$$\log k = -y/x \log [E] + y/x \log C + \text{Constant} \tag{5.19}$$

This last equation is very fundamental for ion chromatography. It predicts a linear decrease in log k as the eluent concentration (E) is increased. At a fixed E concentration, it also predicts a linear increase in log k as the resin capacity is increased.

5.3 Selectivity of Sulfonated Cation-Exchange Resin for Metal Cations

The selectivity of sulfonated ion-exchange resins for metal cations is often expressed qualitatively in terms of elution orders. Numerical selectivity data for cations are limited [3–5]. Strelow and co-workers [6–8] published comprehensive lists of distribution coefficients for metal ions with perchloric acid and other mineral acid eluents. However, their data are for sulfonated gel resins of high exchange capacity.

Sevenich and Fritz [9] published a comprehensive study of metal cation selectivity with resins of a more modern type. The studies were made on a column packed with a 12 % cross-linked polystyrene–divinylbenzene resin 12–15 mm in diameter and with an exchange capacity of 6.1 µequiv/g. The resins were prepared by rapid sulfonation so that the sulfonic acid groups are concentrated on the outer perimeter of the resin beads [10].

5.3.1 Elution with Perchloric Acid and Sodium Perchlorate

The approach is to measure retention times of metal cations on columns containing low-capacity resins with eluents containing perchloric acid or various perchlorate salts. The perchlorate anion is used to eliminate any possible complexing of a metal ion by the eluent anion. Retention factors (k) are calculated from the retention data.

Eluents containing hydrogen ions or sodium ions were used at concentrations ranging from 0.10 to 1.0 M. In general, narrow chromatographic peaks were obtained, although the rare earth peaks were broad with some tailing.

The retention factors were calculated from the retention times and are given in Tables 5.4 and 5.5. The data show that, as expected, eluents containing sodium(I) are more efficient than those containing hydrogen(I). The data in these tables are arranged in order of increasing capacity factors. In this way it is possible to compare the relative affinities of the various divalent and trivalent metal ions for resin sites.

Note that there are crossovers in the retention factors of lead(II) and several trivalent metal ions as the eluent concentration is increased.

The elution order of metal ions reported in Table 5.5 is similar to that reported earlier by Strelow and Sondorp [8] with perchloric acid eluent and gel resins of high exchange capacity. Strelow and co-workers noted anion-complexing effects on elution orders in several cases when eluent acids other than perchloric acid were used.

Eq. 5.19 predicts that a plot of k against the activity of the eluent ion should be a straight line with a slope of $-y/x$. If the eluent ion is monovalent, the slope should be $-y$, where y is the charge on the metal ion, M.

The data in Table 5.4 (perchloric acid eluent) and Table 5.5 (sodium perchlorate eluent) were plotted according to Eq. 5.19. Linear plots were obtained in all cases. (See Fig. 1 for an example.) The slopes, obtained by linear regression, are given in Table 5.4 for perchloric acid and in Table 5.5 for sodium perchlorate. In most cases, the negative slope was very close to the charge on the metal ion. The somewhat low values obtained for Be(II) and Bi(III) in Table 5.6 may be due to partial hydrolysis of

Table 5.4. Retention factors (k) for various cations when perchloric acid eluents are used.

Cation	Perchloric acid concentration (M)							
	1.00	0.85	0.75	0.60	0.50	0.40	0.25	0.10
V(IV)	0.08	0.10	0.12	0.16	0.28	0.40	0.91	5.23
Be(II)	0.10	0.18	0.24	0.30	0.38	–	1.86	4.59
Zr(IV)	0.14	0.20	0.24	0.38	0.65	0.95	2.00	13.31
Mg(II)	0.16	0.18	0.26	0.36	0.51	0.67	1.39	7.49
Fe(II)	0.16	0.18	0.26	0.38	0.53	0.85	1.94	9.19
Mn(II)	0.16	0.18	0.26	0.42	0.55	0.95	2.02	12.14
Hg(II)	0.16	0.18	0.30	0.42	0.69	0.99	2.99	11.70
Ni(II)	0.16	0.22	0.28	0.38	0.59	0.89	2.00	12.75
Zn(II)	0.12	0.18	0.30	0.46	0.71	0.97	2.55	12.67
Cu(II)	0.16	0.22	0.32	0.52	0.61	0.97	3.19	12.83
U(VI)	0.16	0.22	0.28	0.40	0.67	0.99	3.52	12.83
Co(II)	0.20	0.22	0.28	0.38	0.48	1.05	2.02	13.56
Cd(II)	0.22	0.28	0.32	0.59	0.77	1.39	3.56	25.25
Ca(II)	0.32	0.40	0.57	0.97	1.47	2.26	5.98	21.41
Sr(II)	0.44	0.66	0.81	1.39	2.06	3.09	8.55	35.35
Ba(II)	1.01	1.29	1.52	2.73	3.94	6.63	15.76	–
Pb(II)	1.21	1.64	1.90	2.99	4.75	7.37	28.48	100.0
Al(III)	0.68	1.01	1.60	2.95	5.27	9.90	43.0	–
Fe(III)	1.07	1.68	2.89	5.54	9.98	18.42	80.0	–
V(III)	1.13	2.00	2.89	4.67	10.10	20.08	–	–
In(III)	1.15	2.12	3.05	5.84	10.59	19.84	63.8	–
Lu(III)	2.67	4.71	6.57	13.07	27.1	50.5	–	–
Yb(III)	2.93	5.15	7.17	14.14	28.7	52.5	–	–
Tm(III)	2.97	5.25	7.54	14.75	28.9	57.0	–	–
Y(III)	3.13	5.94	8.18	16.38	30.1	59.6	–	–
Er(III)	3.13	5.31	7.92	15.72	30.5	59.8	–	–
Ho(III)	3.37	5.62	8.59	16.71	32.9	63.4	–	–
Dy(III)	3.78	6.10	9.25	18.61	35.8	70.9	–	–
Gd(III)	4.89	8.40	12.48	24.65	47.7	91.3	–	–
Eu(III)	5.45	9.25	13.84	27.47	52.9	105.0	–	–
Sm((III)	6.04	11.37	15.43	30.30	59.4	116.0	–	–
Nd(III)	7.13	13.64	17.76	35.35	69.3	137.8	–	–
Pr(III)	7.44	14.22	18.57	37.58	71.3	142.2	–	–
Ce(III)	8.69	14.44	21.2	43.2	80.8	167.7	–	–
La(III)	9.68	15.64	23.8	47.9	93.7	187.7	–	–
Bi(III)	8.75	13.1	21.2	31.9	53.7	135.0	–	–

Table 5.5. Retention factors (k) of various cations when sodium perchlorate eluents are used.

Cation	Sodium perchlorate concentration (M)				
	1.00	0.75	0.50	0.25	0.20
Be(II)	0.06	0.06	0.12	–	1.05
Mg(II)	0.06	0.10	0.18	0.73	1.23
Ni(II)	0.06	0.14	0.26	1.09	2.42
Fe(II)	0.08	0.14	0.32	1.15	3.58
Co(II)	0.08	0.16	0.20	1.25	2.79
Zn(II)	0.10	0.12	0.30	1.03	2.20
Hg(II)	0.10	0.18	0.36	1.11	2.95
Zr(IV)	0.12	–	0.30	1.29	2.71
V(IV)	0.14	0.16	1.37	0.95	2.40
Mn(II)	0.16	0.20	6.26	1.21	3.45
Cd(II)	0.16	0.22	0.67	1.49	3.23
Ca(II)	0.18	0.30	0.48	2.06	4.36
Cu(II)	0.20	0.30	0.46	1.31	2.79
Sr(II)	0.28	0.51	0.97	3.62	5.92
U(VI)	0.61	0.87	1.27	3.54	7.25
Ba(II)	0.75	1.21	2.26	4.46	–
Al(III)	–	0.75	2.32	–	–
Pb(II)	1.15	1.51	3.03	–	27.7
Lu(III)	1.31	2.63	6.91	44.24	–
Yb(III)	1.37	2.87	7.29	62.6	–
Tm(III)	1.56	3.03	7.92	64.8	–
Y(III)	1.54	2.95	8.28	65.0	–
Er(III)	1.62	3.03	8.34	67.1	–
Ho(III)	1.66	3.31	8.95	67.9	–
Dy(III)	1.94	3.98	10.85	86.1	–
Tb(III)	2.19	4.12	11.33	88.9	–
Gd(III)	2.51	4.89	13.19	104.0	–
Eu(III)	2.79	5.49	14.71	116.2	–
Sm(III)	3.29	6.00	16.55	123.6	–
Nd(III)	3.13	6.24	16.85	131.7	–
Pr(III)	3.21	6.20	17.92	138.6	–
Ce(III)	3.33	6.83	20.40	147.7	–
La(III)	3.80	7.86	21.42	173.5	–
In(III)	3.33	5.01	–	–	–

Table 5.6. Linear regression data for log k plotted against log $[HClO_4]$.

Cation	Slope (this work)	γ (Ref.*)	
Be(II)	–1.64	–1.02	–0.990
Mg(II)	–1.66	–1.25	–0.998
Ca(II)	–1.87	–1.54	–0.996
Sr(II)	–1.91	–1.66	–0.998
Ba(II)	–1.97	–1.92	–0.997
Mn(II)	–1.89	–1.36	–0.998
Zn(II)	–1.98	–1.32	–0.998
Ni(II)	–1.87	–1.35	–0.998
U(VI)	–1.96	–1.32	–0.996
Cu(II)	–1.92	–1.33	–0.997
Fe(II)	–1.79	–1.36	–0.999
Co(II)	–1.86	–1.34	–0.992
Cd(II)	–2.08	–1.40	–0.997
V(IV)	–1.83	–1.11	–0.997
Pb(II)	–2.00	–1.88	–0.994
Hg(II)	–1.92	–1.76	–0.997
Zr(IV)	–1.95	–	–0.999
Al(III)	–2.88	–1.80	–0.999
Bi(III)	–2.67	–2.24	–0.992
Fe(III)	–2.99	–1.81	–1.000
In(III)	–2.78	–1.87	–0.999
V(III)	–2.87	–	–0.996
Lu(III)	–3.00	–	–0.998
Yb(III)	–2.95	–	–0.998
Tm(III)	–2.99	–	–0.999
Y(III)	–2.95	–2.20	–0.999
Er(III)	–3.00	–	–0.999
Ho(III)	–3.01	–	–0.999
Dy(III)	–3.01	–2.23	–0.999
Tb(III)	–3.04	–	–0.998
Gd(III)	–2.97	–	–0.999
Eu(III)	–3.01	–	–0.999
Sm(III)	–2.97	–	–0.999
Nd(III)	–2.97	–	–0.998
Pr(III)	–2.95	–	–0.998
Ce(III)	–3.01	–2.48	–0.999
La(III)	–3.04	–2.49	–0.999

* Calculated from data of Strelow and Sondorp [8].

the metal cation. The slopes obtained for vanadium(IV) and zirconium(IV) indicate that the metal ions are present as VO^{2+} and ZrO^{2+}, respectively.

The necessity for using the activity of the eluent should be emphasized. When the concentration of H^+ (in $HClO_4$) was used, the slopes for the rare earths were approximately –3.25. However, the slopes obtained using the activity were very close to the theoretical value of –3.0.

In view of the very low capacity of the resin used, it may seem remarkable that slopes so close to the theoretical values were obtained. One might expect the ion-exchange sites to be so scattered that it would be impossible for a 3+ cation to exchange with three sites. However, other work has demonstrated that the exchange sites are in a thin layer on the outer perimeter of the ion-exchange bed. The results obtained here indicate that the concentration of sites in this outer layer is sufficiently dense that essentially theoretical exchange is obtained for polyvalent metal ions.

5.3.2 Elution with Divalent Cations

Metal cations were eluted with eluent containing magnesium perchlorate. The divalent magnesium cation is a more effective eluent than H^+ or Na^+, so lower concentrations are needed. In the range 0.5 mM to 10 mM Mg^{2+}, the linear plots of log k vs. log $[Mg^{2+}]$ for several divalent metal cations gave a slope close to the theoretical value of –1.0 (Table 5.7).

Several divalent cations were also eluted with eluent containing the ethylenediammonium- or *m*-phenylenediammonium 2+ cation. Capacity factors for these and other eluent are summarized in Table 5.8. It can be seen that the 2+ diamine cations are much more efficient than Mg^{2+}, Na^+, or H^+.

5.3.3 Effect of Resin Capacity

The capacity factors for metal cations were measured with 0.75 M perchloric acid eluent on resins of 6.1, 13.8, and 24.9 μequiv/g exchange capacity. The results, shown in Table 5.9, show a substantial increase in capacity factors with increased resin capacity. According to Eq. 5.19, the logarithm of k should vary linearly with the logarithm of resin capacity, C. Linear plots are indeed obtained but the slopes are lower than the theoretical values and they vary from one metal ion to another.

An explanation of these observations might be that sulfonation of the resins proceeds inwardly from the outside of the resin beads. Thus, resins of higher capacity are apt to have thicker sulfonation layers. The microscopic selectivity coefficient may well change in various parts of the sulfonation layer. The selectivity coefficients measured would be an average of different microscopic values and might not follow the behavior predicted by Eq. 5.19.

Table 5.7. Linear regression data for log k plotted against log [NaClO$_4$].

Cation	Slope	Correlation coefficient
Be(II)	−2.06	−0.975
Mg(II)	−2.04	−0.998
Ca(II)	−2.11	−0.991
Sr(II)	−2.05	−0.999
Ba(II)	−1.75	−1.000
Ni(II)	−2.38	−0.995
Fe(II)	−2.44	−0.989
Co(II)	−2.34	−0.983
Zn(II)	−2.12	−0.991
V(IV)	−1.95	−0.952
Hg(II)	−2.15	−0.990
Zr(IV)	−2.10	−0.998
Mn(II)	−2.03	−0.960
Cd(II)	−1.99	−0.989
Cu(II)	−1.70	−0.987
U(VI)	−1.62	−0.986
Pb(II)	−2.24	−0.990
In(III)	−3.09	−0.998
Lu(III)	−2.80	−0.999
Yb(III)	−3.03	−0.997
Tm(III)	−2.98	−0.997
Y(III)	−3.00	−0.998
Er(III)	−2.99	−0.997
Ho(III)	−2.96	−0.999
Dy(III)	−3.02	−0.998
Tb(III)	−2.98	−0.998
Gd(III)	−2.98	−0.998
Eu(III)	−2.98	−0.998
Sm(III)	−2.94	−0.998
Eu(III)	−2.98	−0.998
Sm(III)	−2.94	−0.998
Nd(III)	−2.97	−0.998
Pr(III)	−3.00	−0.998
Ce(III)	−3.02	−0.999
La(III)	−3.04	−0.998

Table 5.8. Comparison of capacity factors (k) of various cations with cationic eluents. (Concentrations: $HClO_4$, 0.250 M; $NaClO_4$, 0.100 M; all others are 1.0×10^{-3} M, pH 2.5; t_o = 0.495 min. EnH_2^{2+} = ethylenediamine eluent; $PhenH_2^{2+}$ = *m*-phenylenediamine eluent.

Cation	Eluent				
	$HClO_4$	$NaClO_4$	EnH_2^{2+}	$PhenH_2^{2+}$	$Mg(ClO_4)_2$
Ba(II)	15.8	–	14.00	17.8	–
Be(II)	1.86	1.05	1.58	–	–
Ca(II)	5.98	4.36	4.81	5.21	–
Cd(II)	3.56	3.23	3.27	3.72	15.8
Co(I)	2.02	2.79	2.97	2.97	14.2
Cu(II)	3.19	2.79	2.46	3.54	15.8
Fe(II)	1.94	3.58	2.87	2.85	13.6
Hg(II)	2.00	2.95	2.77	3.15	14.6
Mg(II)	1.39	1.23	1.98	2.40	–
Mn(II)	2.02	3.45	3.01	3.01	15.4
Ni(II)	2.02	2.42	2.95	3.01	14.0
Pb(II)	28.5	27.7	19.8	24.2	–
Sr(II)	2.48	5.92	6.42	7.27	–
U(VI)	3.52	7.95	6.24	6.02	22.0
V(IV)	0.91	0.95	2.10	2.20	7.4
Zn(II)	2.55	2.20	2.75	2.91	13.1
Zr(IV)	2.00	2.71	2.85	3.09	14.4

Table 5.9. Retention factors (k) for cations with resins of different bulk densities. The eluent is 0.75 M perchloric acid.

Cation	Resin capacity (μequiv/g)		
	6.1	13.8	24.9
V(IV)	0.12	0.40	0.81
Be(II)	0.24	0.32	0.38
Zr(IV)	0.24	0.40	1.03
Mg(II)	0.26	0.32	0.65
Fe(II)	0.28	0.40	0.93
NI(II)	0.28	0.51	0.87
U(VI)	0.28	0.59	1.21
Mn(II)	0.28	0.67	1.27
Zn(II)	0.30	0.42	0.85
Cu(II)	0.32	0.46	0.95
Cd(II)	0.32	0.51	1.11
Ca(II)	0.57	1.05	2.46
Ba(II)	1.52	5.05	–
Pb(II)	1.90	4.97	11.05
Fe(III)	2.89	5.68	11.0
In(III)	3.05	6.12	–
Lu(III)	6.6	15.9	35.4
Yb(III)	7.2	16.4	36.8
Tm(III)	7.5	17.8	38.4
Er(III)	7.9	18.8	42.4
Y(III)	8.2	19.3	42.8
Ho(III)	8.6	20.4	46.9
Dy(III)	9.3	24.4	55.4
Tb(III)	10.6	26.7	65.0
Gd(III)	12.5	31.5	73.1
Eu(III)	13.8	35.4	84.2
Sm(III)	15.4	39.4	95.8
Nd(III)	17.8	45.0	106.3
Pr(III)	18.6	48.9	115.0
Ce(III)	21.2	56.6	124.6
La(III)	23.8	68.1	172.1
Bi(III)	–	42.4	122.6

5.4 Separation of Divalent Metal Ions with a Complexing Eluent

The cation separations discussed thus far have been possible because of differing affinities of the various ions for the cation-exchange resin. In traditional ion-exchange chromatography, the number of metal cations that can be separated in this manner has always been limited. The bulk of successful separations of metal cations has been made possible by using an eluent of selective complexing ability. For this reason, the possibility of using an auxiliary complexing reagent in the eluent has now been considered for cation chromatography with conductivity detection [11].

In Chapter 7, a number of separations are described in which a weakly complexing ligand is added to the eluent to partially complex the sample cations. Separation of metal ions may now be based on differences in partial complexation as well as on differences in the selectivity of the ion exchanger for free metal ions.

5.4.1 Principles

Addition of a weak complexing anion (L), such as tartrate or α-hydroxyisobutyrate (HIBA), will convert part of a metal ion to a complexed form that is uncharged or has a lower positive charge than the free metal ion. This will cause the metal ion to elute more rapidly. Referring to Eq. 5.20, the metal ion in solution (I_{soln}) will be partially complexed:

$$I + L \rightleftarrows IL + IL_2 + \dots \tag{5.20}$$

The ratio of free metal ion (I) to the total amount that is present in solution has been defined by α.

$$\alpha = \frac{[I]}{[I]+[IL]+[IL_2]+\dots} \quad \frac{I}{I'} \tag{5.21}$$

For convenience, charges are omitted from the ionic species. The quantity, (I_{soln}) in Eq. 5.21 may now be represented by $I'\alpha$. Continuing the derivation given earlier (Section 5.2.4), Eq. 5.17 becomes:

$$k^x = \alpha^x C^y K/[E_{soln}]^y \tag{5.22}$$

and Eq. 5.19 now has an additional term involving α.

$$\log k = \log \alpha_1 - y/x \log [E_{soln}] + y/x \log C + \text{Constant} \tag{5.23}$$

The value of α_I will depend on the formation constants of the metal ion–ligand formation constants, the concentration of excess ligand in the mobile phase and the pH. Calculation of α_I according to Ringbom [12] follows the equation:

$$1/\alpha_I = 1 + B_1[L] + B_2[L]^2 + B_3[L]^3 + ...$$ (5.24)

where B_1, B_2, B_3 are complex formation constants that are available in the literature for metal complexes of various ligands, and [L] is the concentration of the free ligand anion in solution.

The linear dependence of log k on log α_I has been demonstrated for the ion chromatographic behavior of several lanthanide 3+ cations, with an eluent containing tartrate (see Fig. 5.2). Increasing values of α_I mean that more of the lanthanide exists as the 3+ cation and less is complexed by the ligand.

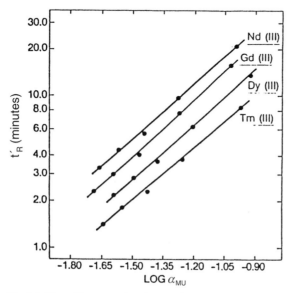

Fig. 5.2. Plot of log retention time against log α_M for elution of rare earth cations with ethylenediammonium tartrate (*Courtesy of G. J. Sevenich*).

References

[1] A. Rahman and N. E. Hoffman, Retention of organic cations in ion exchange chromatography. *J. Chromatogr. Sci.*, 28, 157, 1990.
[2] K. Lee and N. E. Hoffman, Retention of some simple organic ions in ion exchange HPLC, *J. Chromatogr. Sci.*, 30, 98, 1992.
[3] O. D. Bonner and L. L. Smith, A selectivity scale for some bivalent cations on Dowex 50, *J. Phys. Chem.*, 61, 326, 1957.
[4] J. F. Helfferich, *Ion Exchange*, McGraw-Hill, New York, 1962, p. 169.
[5] W. Rieman III and H. F. Walton, *Ion Exchange in Analytical Chemistry*, Pergamon, Oxford, 1970, p. 45.
[6] F. W. E. Strelow, An ion exchange selectivity scale of cations based on equilibrium distribution coefficients., *Anal. Chem.*, 32, 1185, 1960.
[7] F. W. E. Strelow, R. Rethemeyer and C. J. C. Bothma, Ion exchange selectivity scales for cations in nitric acid and sulfuric acid media with a sulfonated polystyrene resin, *Anal. Chem.*, 37, 106, 1965.
[8] F. W. E. Strelow and H. Sondorp, Distribution coefficients and cation-exchange selectivities of elements with AG50W-X8 resins in perchloric acid, *Talanta*, 19, 1113, 1972.

[9]   G. J. Sevenich and J. S. Fritz, Metal ion selectivity on sulfonated cation-exchange resins of low capacity, *J. Chromatogr.*, 371, 361, 1986.
[10] G. J. Sevenich and J. S. Fritz, Preparation of sulfonated gel resins for use in ion chromatography, Reactive Polymers, 4, 195, 1986.
[11] G. J. Sevenich and J. S. Fritz, Addition of complexing agents in ion chromatography for separation of polyvalent metal ions, *Anal. Chem.*, 55, 12, 1983.
[12] A. Ringbom, "Complexation in Analytical Chemistry:", *Interscience*, New York, 1963.

6 Anion Chromatography

6.1 Scope and Conditions for Separation

A very large number of anions have been separated successfully by anion chromatography. Table 6.1 lists many of these anions, grouped according to their general type. Particular attention has been paid to the so-called common anions. These can be separated in a single run with suppressed conductivity detection (Section 6.2), non-suppressed conductivity (Section 6.3), or by one of the other detection methods discussed in later sections. If necessary, the sample is diluted so that the ions to be determined are in the low to medium concentration range. However, sample anions down into the lower parts-per-billion (ppb) range can be successfully separated and quantified.

Table 6.1. Anions determined by ion chromatography (Grouped according to general type).

1. Common anions: Fluoride, bromide, nitrite, nitrate, phosphate, sulfate.

2. Polarizable: Iodide, thiocyanate, thiosulfate, chromate, molybdate, tungstate.

3. Inorganic anions of weak acids: Borate, bicarbonate, carbonate, cyanide, silicate.

4. Other inorganic anions: Arsenite, arsenate, azido (N_3^-), bromate, chlorite, cyanate, chlorate, perchlorate, iodate, periodate, sulfamate ($NH_2SO_3^-$) sulfite, selenite, selenate.

5. Smaller organic anions: Amino acids, alkane carboxylic acids (formate, acetate, propionate, butyrate), chloro carboxylic acids (chloroacetate, dichloroacetate), hydroxy acids (hydroxyacetate, lactate, tartrate, citrate), glycolate, gluconate, pyruvate, dicarboxylic acids (oxalate, malonate, succinate, glutarate, fumarate, maleate), alkanesulfonic acids (methanesulfonate, ethanesulfonate).

6. Larger organic anions: aromatic carboxylic acids, aromatic sulfonic acids, carbohydrates (aldoses, ketoses), nucleotides, nucleic acids, proteins.

Polarizable anions include those listed in Group 2 of the table. These anions have a relatively high affinity for the ion-exchange stationary phase and therefore require a stronger eluent for their separation.

The anions listed in Group 3 require an alkaline pH for their separation so that they will be in the anionic rather than the molecular form. Some of the ions listed in Group 4 will also require an alkaline eluent for separation.

Smaller organic anions such as those listed in Group 5 are usually sufficiently hydrophilic to be separated in aqueous solution. If necessary, some methanol or acetonitrile may be included in the eluent to increase their solubility and also to avoid anions for the solid phase. Although the anion-exchange resins in Table 6.2 were designed specifically for use with suppressed conductivity detection (Section 6.2), they are also generally useful for IC involving other modes of detection.

The anion-exchange described in Tables 6.4 and 6.5 are not latex-coated but their exchange groups arc mostly in a narrow band near the outside perimeter of the spherical ion-exchange particle. In some instances the ion-exchange part of the particle consists of a functionalized polymer grafted onto the outer surfaces of the substrate. It should be emphasized that these columns work well for all forms of detection including suppressed conductivity. But the properties of these columns vary, so again it is important to select a column that is compatible with the intended IC separation.

6.1.2 Separation Conditions

The goals of the ion chromatographer are first, to achieve a satisfactory separation of the sample components of interest and second, to perform the separation as quickly as possible. Several parameters can be manipulated to fulfill these objectives.

Resin structure. Changes in the chemical composition may alter the selectivity. For example, an ethanolamine group instead of the usual methyl groups in the quaternary ammonium function increases the affinity of the OH^- for the ion exchanger and makes NaOH or KOH a more powerful eluent.

Resin capacity. A lower exchange capacity will result in faster elution of analyte ions or will permit elution with an eluent of lower anion concentration. Divalent anions are affected more relative to monovalent anions.

Column diameter. Use of a column with i.d. of 2 mm instead of 4.0 or 4.6 mm will result in a faster elution at the same volume flow rate.

Eluent strength. Anion eluents vary considerably in their ability to elute sample anions. Several anions commonly used for IC with suppressed conductivity detection are listed in Table 6.5. An anion of higher charge, such as carbonate, has greater eluting strength than an eluent containing a monovalent anion.

Eluent concentration. A higher concentration of the eluent anion will result in shorter retention times for sample ions.

Organic solvents. A certain percentage (perhaps 10 to 20 %) of methanol or acetonitrile may be incorporated in the mobile phase to ensure solubilization of sample components. Another function of organic solvents is to improve compatibility of sample ions for ion exchangers with a hydrophobic matrix. Tailing of peaks of polarizable anions may be reduced by incorporation of an organic solvent in the eluent.

Isocratic or gradient elution. Use of a fixed eluent (isocratic elution) is the simplest, most straightforward method for IC. But when there is a large difference between retention times of early- and late-eluting ions or when a very complex sample is to be analyzed, gradient elution may be desirable or even essential. However, it should be kept in mind that after a gradient elution, reequilibration with eluent of the lower strength is necessary before the next run can begin. Quantification with gradient elution is also more complex. It is best to calibrate peak height or area under the exact conditions of the gradient elution.

6.2 Suppressed Anion Chromatography

Chemical suppression provides a simple, yet elegant way to reduce the background conductance of the eluent and at the same time to enhance the conductance of sample ions. In its original form, a second ion-exchange column was placed between the separator column and the conductivity cell. For anion analysis a basic anion was used in the eluent and a large, H^+-form cation exchange column was used as the second "stripper" column [1].

Modern methods of chemical conductivity suppression utilize much smaller and more efficient devices than the original stripper columns, but the basic principles are largely unchanged. The reactions for suppressed anion chromatography are as follows:

Eluent

NaOH	+	RSO_3H	→	H_2O	+	RSO_3Na
Eluent		Suppressor		Suppressed Eluent		Suppressor
(Highly Conductive)				(Weakly Conductive)		

Analyte

NaX	+	RSO_3H	→	HX	+	RSO_3Na
Analyte Salt		Suppressor		Acid form of Analyte		Suppressor
(Conductive)				(Highly Conductive)		

The basic eluent (OH^-) is neutralized by H^+ from the cation exchanger of the suppressor to form water. A sample zone passing through the suppressor is converted from the sodium as the counterion (Na^+OH^-) to the more highly conducting hydrogen counterion (H^+X^-).

The major devices for suppressed conductivity detection in ion chromatography have been reviewed [2]. These are described in chronological order in the following sections.

6.2.1 Packed-Bed, 1975 [1]

These had some major drawbacks. To contain enough resin for continuous operation, the suppressors had a very large dead volume that caused considerable peak dispersion and broadening. Regeneration of the resin bed was another serious problem. After several hours of operation, the ion exchange bed became expanded and had to be regenerated. This was done offline with sulfuric acid (for anion chromatography); it was flushed with water, then placed back on-line.

6.2.2 Fiber Suppressors, 1981 [3]

The fiber suppressor was the first device based on the use of an ion-exchange membrane. It consisted of a long, hollow fiber made of a semi-permeable ion-exchange material. Column effluent containing zones of separated sample ions passed through the hollow center of the fiber. Here the sodium counterion was exchanged for H^+ from the membrane. The outside of the hollow fiber was bathed in an acidic solution, allowing for continual replacement of the H^+ as the effluent passed through. The main advantage of this design was that it permitted continuous operation of the IC system. Band broadening in this suppressor was less than with the large packed-bed devices but was still significant. Fiber suppressors were also limited in their ability to suppress flow rates above 2 mL/min or eluents above 5 mM concentration.

6.2.3 Membrane Suppressors, 1985 [4]

A flat membrane suppressor from Dionex, known as the Micro-Membrane Suppressor (MMS) had a much higher capacity and lower dead volume than previous devices and was able to operate around the clock with minimal attention. The internal design of the MMS is shown in Fig. 6.1. Two semi-permeable ion-exchange membranes are sandwiched between three sets of ion-exchange screens. The eluent screen is of fine mesh to promote the suppression reaction while occupying a very low volume. The ion-exchange membranes on either side of this screen define the eluent chamber. There are two ion-exchange regenerant screens that permit tortuous flow of

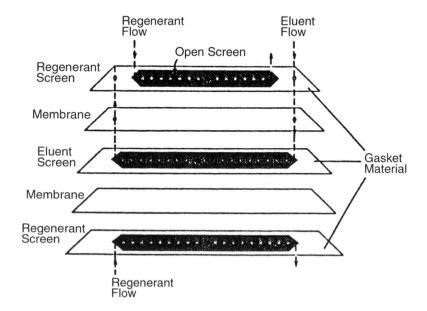

Figure 6.1. Internal design of the MicroMembrane Suppressor. (From Ref. [2] with permission).

the regenerant solution towards the membranes. These screens provide a reservoir for suppressing ions without having a counterion present.

The flow pattern for the anion suppressor is shown in Fig. 6.2. Column effluent flowing through the suppressor exchanges Na^+ for H^+ from the cation exchange membranes, as shown in the middle part of the figure. Since the suppressor is actually a sandwich configuration with fairly broad cation-exchange membranes placed very close together, the exchange reaction proceeds rapidly and there is adequate exchange capacity to handle eluents of higher concentrations. A mineral acid such as dilute sulfuric acid flows through outer parts of the suppressor to provide continuous regeneration. Regenerant flow is counter-current to the column effluent and at approximately three to ten times the chromatographic flow rate.

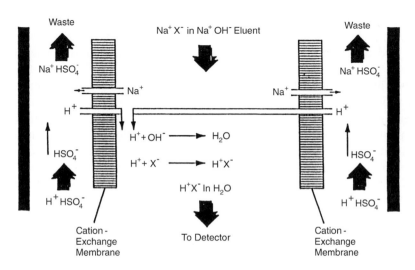

Figure 6.2. Suppression mechanism for the Anion MicroMembrane Suppressor. (From Ref. [2] with permission).

A major drawback of the membrane suppressors was that they required a constant flow of regenerant for continuous suppression. This could consume up to 10 mL/min of regenerant solution to reduce the large volumes of regenerant needed, an accessory for continuous regenerant recycling was introduced by Dionex in 1987. A large ion-exchange cartridge was used to remove the comparatively low concentrations of waste products (Na^+, etc.) and replace it with fresh regenerant ions (H^+). A pump recirculates the regenerant through the suppressor cartridge. The net effect is that only a small reservoir of regenerant solution is required for effective operation.

6.2.4 Electrolytic Suppressors

The ideal way to regenerate a suppressor for IC is to electrolyze water to produce the H^+ or OH^- needed. Strong and Dasgupta invented a practical suppressor of this type in 1989 [5].

In this device, a platinum wire-filled tube made of a Nafion cation-exchange membrane is inserted into another, larger Nafion tube and coiled into a helix. The helical assembly is inserted within an outer jacket packed with granular conductive carbon. An alkaline eluent, for example, NaOH or Na_2CO_3, flows in the annular channel between the two membranes and pure water flows through the inner membrane and the outer jacket countercurrent to the direction of eluent flow. A DC voltage (3 to 8 V) is applied across the carbon bed and the platinum wire. Sodium ions in the eluent migrate to the cathode compartment resulting in water as the suppressed eluent. Up to 500 µL/min of sodium hydroxide could be suppressed effectively with a membrane 50 cm in length. The band dispersion was 106 µL for a 20 µL sample.

In 1992 Dionex introduced a commercial electrochemical suppressor called a Self Regeneration Suppressor, or SRS [6]. The internal design is similar to the membrane suppressor, but the regenerating ion (H^+ for anion chromatography) is produced by electrolysis of water. This allows the use of very low flow rates for regenerant water and avoids the use of independent chemical feed needed for earlier suppression devices.

The mechanism for the Anion SRS is described in Fig. 6.3. Hydrogen ions generated at the anode traverse the cation-exchange membrane to neutralize the basic eluent. Sodium counterions are attracted to the negatively charged cathode, where they permeate the membrane in the cathode chamber and pair off with electrogenerated hydroxide ions to maintain electric neutrality. Waste gases, hydrogen from the cathode and oxygen from the anode, are vented with a liquid waste of aqueous sodium hydroxide. As with other suppressors, analyte ion signals are enhanced by exchange of their counterions for hydrogen ions.

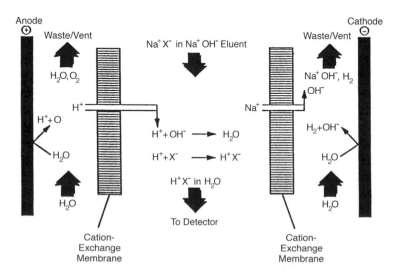

Figure 6.3. Mechanism of suppression for the Anion Self-Regenerating Suppressor. (From Ref. [2] with permission).

6.2.5 Solid-Phase Reagents, 1990 [7]

Gjerde and Benson discovered that post-column addition of a suspension of sulfonated polystyrene particles may be used to reduce the background conductance of basic eluents used in anion chromatography [7]. The eluent cation (typically Na^+) is also replaced in the analyte ion bands by the more highly conducting H^+ as the counterion to a sample anion. Since the added reagent is a solid, it is "invisible" to detectors that respond only to the liquid phase, for example, conductivity and potentiometric detectors.

A typical solid-phase reagent (SPR) is a colloidal polystyrene sulfonic acid material, 2.3 mequiv/g exchange capacity used as a 0.5 to 1.0 % suspension at a flow rate of 1.0 mL/min. This material stays in suspension and undergoes rapid ion exchange with ions in the column effluent. A detailed diagram of the hardware required is shown in Fig. 6.4.

This unique system works well for gradient elution [8] and for the determination of analytes of extreme concentration ratios [9].

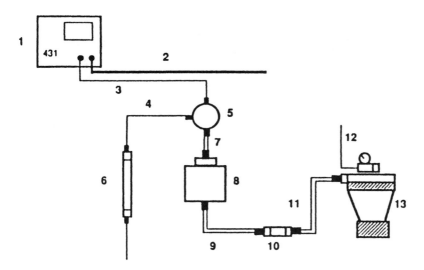

Figure 6.4. Detailed diagram of hardware configuration for post-column addition of SPR. (1 = Conductivity detector: Waters 431 detector, four electrode cell design; 2 = waste line: 4 × 0.009 in. stainless connected to 431 + 24 × 1/16 × 0.060 in PTFE tubing; 3 = tee to 431: 15 × 1/16 × 0.010 in PTFE to 431 inlet; 4 = column to tee: shortest 1/16 × 0.010 in PTFE from column to tee; 5 = tee: Unmount tee from check valve block for shortest path length; 6 = analytical column: Waters IC PAK A or IC PAK A HR; 7 = check valve to tee: 2 × 1/8 in o.d. PTFE; 8 = check valve; 9 = polisher column to check valve: 3 × 1/8 in o.d. PTFE; 10 = polisher column: 8 × 25 mm containing AGI × 8, 200 mesh; 11 = reservoir to polisher column: 12 × 1/8 in. o.d. PTFE; 12 = air supply: minimum of 90 p.s.i. compressed air supply; 13 = reservoir for SPR: reconfigure with outlet on left side. From Ref. [9] with permission.)

6.2.6 Eluents

The major eluents that have been used for suppressed anion chromatography are listed in Table 6.6. These are all basic eluents that produce a suppressor product that is a very weak acid and therefore one that has a very low conductivity. The ultimate eluent in terms of suppressed conductivity detection is the hydroxide ion, which gives water as the suppressor product.

Table 6.6. Eluents for suppressed anion chromatography.

Eluent	Eluent ion	Suppressor product	Elution strength
Sodium tetraborate	$B_4O_7^{2-}$	H_3BO_3	Very weak
Sodium hydroxide	OH^-	H_2O	Weak
Sodium bicarbonate	HCO_3^-	H_2CO_3	Weak
Sodium bicarbonate/carbonate	HCO_3^-/CO_2^{2-}	H_2CO_3	Medium
Amino acid/NaOH	$RCH(NH_2)CO_2^-$	$RCH(NH_3)CO_2^-$	Medium
Sodium carbonate	CO_3^{2-}	H_2CO_3	Strong

A mixture of sodium bicarbonate and sodium carbonate has been widely used as an eluent for many years. Carbonate with a 2– charge is a stronger eluent than bicarbonate. By using a mixture of the two, the eluent strength can be adjusted as desired.

With the advent of anion exchangers with an increased affinity for hydroxide and suppressors that tolerate a higher eluent concentration, the use of sodium or potassium hydroxide has become more popular. However, it is difficult to remove all of the carbonate from chemical solutions of sodium hydroxide. Electrolytic generation is now the preferred way to produce hydroxide eluents for IC. The product is almost entirely free of carbonate and the electrolytic generation provides excellent control of the concentration. Electrolytic generators are described in Chapter 1.

6.2.7 Typical Separations

Particular attention has been paid to separation of common anions (see Table 6.1) because they are present in so many types of samples. Complete resolution of common anions may take approximately 20 min (Dionex AS9HC, 9.0 mM sodium carbonate), but the use of a column of lower exchange capacity or the use of a 2.0 mm i.d. column can reduce the separation time to approximately 10 min or less. For high throughput of simple, well-characterized samples, use of a column with unusually low exchange capacity will give a good separation of seven common anions in a approximately 2 min (Fig. 6.5).

Separation of other inorganic anion and small organic anions is also possible. Figure 6.6 shows a separation of selenite and selenate as well as several of the common anions. It is often desirable to distinguish between different species of the same element, as was done in this case. Selenium is an essential nutrient but at slightly higher

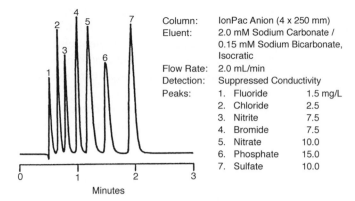

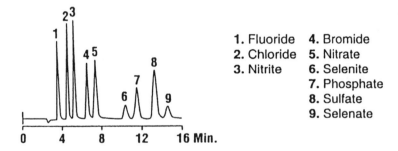

Figure 6.5. Rapid separation of common anions (Courtesy Dionex Corp).

Figure 6.6. Separation of selenite and selenate (Courtesy Alltech).

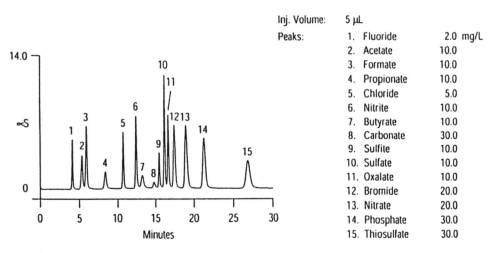

Figure 6.7. Separation of organic and inorganic anions on a 2-mm AS15 column; a potassium hydroxide gradient is used: 9 mM from 0 to 6 min, 9 mM to 47 mM from 6 to 15 min (Courtesy Dionex Corp).

concentrations it can be hazardous. This separation demonstrates that it is not essential to use latex-coated resin for separations with suppressed-conductivity detection.

Samples with a larger number of ions to determine may require gradient elution. A separation of 15 inorganic and organic anions with an electrically generated potassium hydroxide gradient is shown in Fig. 6.7. A 2-mm column with a rather high exchange capacity (AS 15) was used with a 5 µL sample injection.

A large sample volume may be used to achieve very low limits of detection. The chromatogram in Fig. 6.8 was obtained with a column and gradient similar to that in Fig. 6.8, but a 1000 µL sample injection was used to detect sample anions in the very low ppb range.

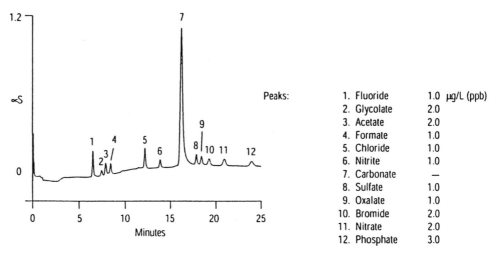

Figure 6.8. Ion separation at very low concentrations with a 1000 µL injection and a potassium hydroxide gradient: 8 mM from 0 to 6 min, 8 mM to 45 mM from 6 to 17 min (Courtesy Dionex Corp.).

6.3 Non-Suppressed Ion Chromatography

6.3.1 Principles

Ion chromatography with suppressed conductivity detection has been extremely successful in filling a large gap that previously existed in inorganic analysis. However, the necessity for a suppressor device does add to the complexity of the instrumentation. It also restricts the type of eluent that can be used and to some extent limits the separating ability of the method.

In the middle to late 1970s several workers experimented with changing the capacity of ion-exchange resins [10]. It was found that a significant reduction in resin ion-exchange capacity will permit a substantial reduction in eluent ion concentration while maintaining retention times of analyte ions in a desirable range. This led to the development of a very simple form of ion-exchange chromatography with conductivity detection [11,12]. At the time it was called single-column ion chromatography (SCIC) and is now referred to as non-suppressed ion chromatography (NSIC).

Non-suppressed ion chromatography employs a conventional liquid chromatographic system with a conductivity detector cell connected directly to the outlet end of an ion-exchange separation column. No suppressor unit is required. The successful development of this method was made possible by three principal innovations: (1) the use of an anion- or cation-exchange resin of very low capacity (initially 0.007 to 0.04 mequiv/g), (2) an eluent with a low ionic concentration and hence a low conductivity, and (3) an eluting ion in the eluent that has a significantly lower equivalent conductance than the analyte ions.

For separation of anions, the ionic concentration of the eluent was typically 0.5 mM in the original work and is seldom more than 5 mM in more recent IC methods. A solution containing the alkali metal salt of benzoic- or phthalic acid is a suitable eluent. The benzoate anion has a limiting equivalent conductance of 32 (S cm^2 equiv^{-1}). By contrast, the equivalent conductance of common anions such as chloride, bromide, nitrate and sulfate is 70 to 80 (S cm^2 equiv^{-1}). Thus when one of these analyte anions passes through the detector cell, there is a significant increase in conductance over the background conductance. Figure 6.9 shows an early separation of chloride, nitrate and sulfate when 0.5 mM potassium phthalate at pH 6.2 is the eluent [10]; the sulfate peaks are a composite of five runs with sulfate concentrations ranging from 2.75 to 13.75 ppm. A plot of sulfate peak height or area vs. concentration is linear.

A separation of eight anions with a more contemporary IC system is shown in Fig. 6.10. Quantitative results are possible in the low-to-medium ppm concentration range for each of the anions.

6.3.2 Explanation of Chromatographic Peaks

When a sample containing salts of various anions, M^+A^-, M^+B^-, M^+C^-, etc. is injected, the sample anions will be taken up by the resin and exchanged for an equivalent amount of eluent anion, E^-. The sample volume is rather small (usually 100 μL). The sample zone travels down the separator column at a rate equal to the eluent flow rate. This zone contains the cations present in the original sample and an eluent anion concentration equivalent to that of the sample anions. If the conductance of the cations and anions in this zone is greater than that of the eluent, a positive pseudo peak will be observed when this zone passes through the conductivity detector. However, if the conductance of ions in this zone is lower than that of the eluent, a negative pseudo peak will result. The pseudo peak in Fig. 6.9 is negative while that in Fig. 6.10 is first negative and then positive. This matrix peak causes no difficulty, provided elution of the sample anions is delayed until after the matrix peak has passed.

After the sample plus has passed, the baseline is quickly restored to that obtained with the eluent alone. The solute anions gradually move down the column with the eluent. The total ion concentration in solution in the column is fixed by the eluent anion concentration because solute anion can only enter the solution phase by uptake of an equivalent number of eluent anions. The change in conductance when a sample solute band passes through the detector results from replacement of some of the eluent anions by solute anions, although the total ion concentration remains constant.

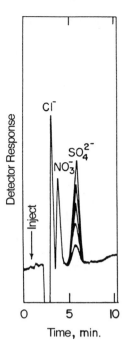

Figure 6.9. Separation of standard solutions of sulfate (2.75–13.75 ppm) from chloride and nitrate. Resin: XAD-1, 44–57 μm 0.04 mequiv/g; eluent, 5.0×10^{-4} M potassium phthalate, pH 6.2.

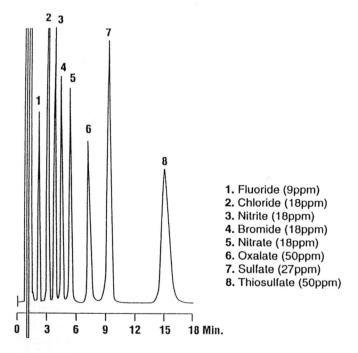

1. Fluoride (9ppm)
2. Chloride (18ppm)
3. Nitrite (18ppm)
4. Bromide (18ppm)
5. Nitrate (18ppm)
6. Oxalate (50ppm)
7. Sulfate (27ppm)
8. Thiosulfate (50ppm)

Figure 6.10. Separation of oxalate, thiosulfate and other anions. Conditions: Allsep anion column 100 × 4.6 mm, 4 mM phthalic acid, pH 4.2, 1.0 mL/min at 35 °C, conductivity detector (Courtesy Alltech).

This change is directly proportional to the sample concentration and the difference in equivalent conductance of the eluent anion and the sample anion. Conductivity detector response is discussed in greater detail in Section 6.3.6.

6.3.3 Eluent

6.3.3.1 General Considerations

The eluent must be carefully chosen if a conductivity detector is to be directly coupled to an anion-exchange separation column. In general, a good eluent is an aromatic organic anion which has a high selectivity coefficient for the anion exchange resin.

As the affinity of the anion exchanger for an eluent anion increases, the eluent concentration needed to move sample anions along the ion-exchange column becomes progressively lower. A low eluent concentration is desirable because of the correspondingly low background conductance. There is probably some limit as to how great the affinity of the ion-exchange resin for the eluent anion should be; if the affinity becomes too great sorption of the eluent anion (as opposed to ion exchange) is likely to occur.

Hence, the major general criterion for eluent selection is a different equivalent conductance compared to the sample ions and a high enough affinity for the resin to promote effective elution of the sample ions. So far we have emphasized that the eluent should have a low equivalent conductance. But an eluent such as sodium hydroxide, with a high equivalent conductance, can be used in non-suppressed IC. This eluent is discussed in Section 6.3.3.5.

Many different eluents have been used for NSIC. The major types are discussed below.

6.3.3.2 Salts of Carboxylic Acids

Lithium, sodium, potassium, or other salts of benzoic acid, phthalic acid, sulfobenzoic acid, citric acid, and others are useful eluents for anions. These are rather large organic anions that are less mobile than most inorganic anions and therefore have lower equivalent conductances. For example, Table 4.1 shows that the benzoate anion has a limiting equivalent conductance of 32 S cm^2 equiv^{-1}, while chloride, nitrate, sulfate, and other typical sample anions have higher equivalent conductances (approximately 70 S cm^2 equiv^{-1}). If a sodium benzoate eluent is used, the equivalent conductance is the sum of sodium ion (50) and benzoate (32), or 85 S cm^2 equiv^{-1}. The equivalent conductance of an anion is the sum of equivalent conductances of the sodium ion (50) and the anion (70), or 120 S cm^2 equiv^{-1}. On an equivalent basis, this amounts to almost a 50 % increase in conductance.

6.3.3.3 Benzoate and Phthalate Salts

The benzoate salt is one of the two most useful carboxylic acid salts for eluents. The other is phthalate. Benzoate salts are useful for separation of acetate, bicarbonate, fluoride, chloride, nitrite, nitrate, and other early-eluting anions. Divalent anion and other late-eluting anions such as thiocyanate and perchlorate are not eluted effectively by benzoate. The concentration of a benzoate eluent that should be used depends on the type and capacity of anion-exchange resin used, but is typically 0.5 to 5.0 mM.

Potassium phthalate eluents can be conveniently prepared by dissolving potassium acid phthalate in pure water and adjusting the pH to around 6.1 to 7.0 In this pH range, the 2- phthalate anion is the predominant species. Phthalate is a more powerful eluent than benzoate and is used for separation of divalent anions and other late-eluting anions such as iodide, thiocyanate, and perchlorate. Early-eluting anions such as bicarbonate, acetate, and fluoride are usually indistinguishable from the pseudo peak when a phthalate eluent is used. The eluting power of phthalate can be modified by changing the pH of the mobile phase. At pH 8.2 the hydrogen phthalate (HPh) form is primarily present in solution. Around pH 8.5, the ionic form has been converted to the phthalate (Ph^{2-}), which is a more powerful eluent by virtue of its 2– charge.

6.3.3.4 Other Eluent Salts

Alkali metal salts of benzenesulfonic acid are similar to benzoate in their eluting power although benzenesulfonate retains its 1– charge at a lower pH.

In some cases, eluents in mixed ionic forms can elute both weakly and strongly retained sample ions in a single run. Solutions of *p*-hydroxybenzoic acid are a good example of a mixed eluent. At pH 8.5, the carboxyl group is completely ionized. The phenolic group is a weaker acid ($pK_a = 9.3$) and becomes increasingly ionized at more alkaline pH values. Thus, by adjusting the pH, a mixture of 1– and 2– driving ions can be obtained, making it possible to elute ions from fluoride through sulfate in a single run.

6.3.3.5 Basic Eluents

Anions of very weak acids such as arsenite, borate, carbonate, cyanide and silicate exist as anions only in basic solution. It is therefore necessary to use a basic eluent to separate these anions. A solution of sodium hydroxide can be used. Detection with sodium hydroxide is different than with the organic salt and acid eluents. Since the hydroxide ion is more mobile and has a higher equivalent conductance than most other anions, the peaks for the sample anions appear as negative peaks (decreased conductance). However, the peak height (or area) is still a function of the amount of sample anion and the sensitivity is even better than with the more acidic eluents where positive peaks are obtained.

The theory of negative peaks for sample anions is easy to understand. Suppose we use 1.0×10^{-3} M sodium hydroxide as the eluent. The background conductance will be the sum of the sodium and hydroxide conductances and will be relatively high. Injection of a sample will result in the uptake of the anions by the resin column with an equivalent amount of resin hydroxide ion passing into solution. Once the matrix peak is through, the anion concentration of the column effluent will be constant, as fixed by the 1.0×10^{-3} M eluent concentration. When a sample anion, A^-, is eluted from the column and passed through the detector, the eluent hydroxide ion will be decreased because a constant anion concentration must be maintained ($[A^-] + [OH^-] = 1.0 \times 10^{-3}$ M). The equivalent conductances of most anions range from about 30 to 80 S cm^2 equiv^{-1}, while the hydroxide ion has an equivalent conductance of 199. Thus, the conductance will decrease when a sample anion is eluted and the height (or area) of this negative peak will be proportional to the concentration of the anion.

Although sodium hydroxide can be used as an eluent for cyanide, acetate, arsenite, fluoride, and other easily eluted anions, the hydroxide ion is a rather weak eluent for many anions. More recent work has shown that a sodium hydroxide solution containing small amounts of sodium benzoate (in about 1:10 molar ratio) behaves similarly to hydroxide (the peaks are usually still of decreasing conductance), but is a more powerful eluent than sodium hydroxide alone.

6.3.3.6 Carboxylic Acid Eluents

The detection sensitivity in NSIC can be improved markedly if the acidic form of benzoic acid is used, rather than the alkali metal salt, in the eluent. Gjerde and Fritz [13] separated seven different anions with a 1.25×10^{-3} M solution of benzoic acid as the eluent. Benzoic acid eluent provides for excellent separations of easily eluted anions, and the sensitivity is several times better than can be obtained with sodium or potassium benzoate.

From the acid ionization constant, $K_a = 6.25 \times 10^{-5}$, it can be calculated that a 1.25×10^{-3} M solution of benzoic acid is 20 % ionized. Thus this eluent has a hydrogen ion and a benzoate ion concentration each of 2.50×10^{-4} M. Comparison of this benzoic acid solution with an eluent containing 2.50×10^{-4} potassium benzoate shows that the two eluents give very similar retention times for inorganic anions (see Table 6.7).

Elution of a sample anion, A, from the resin column involves a one for one exchange with a benzoate ion.

$$HBz$$
$$\updownarrow$$
$$R\text{-}A^- + H^+ + Bz^- \rightleftharpoons R\text{-}Bz^- + H^+ + A^-$$

Table 6.7. Comparison of anion retention times with benzoic acid and potassium benzoate eluents that have the same benzoate ion concentrations. (From Ref. [13]).

Anion	Adjusted retention time benzoic acid, 1.25×10^{-3} M, pH = 3.65 (min)	Adjusted retention time potassium benzoate, 0.25×10^{-3} M, pH = 6.10 (min)
Acetate	1.16	4.36
Propionate	1.98	6.76
Formate	2.56	– [a]
Fluoride	3.62	4.26
Phosphate	4.50	– [a]
Chloride	5.36	5.44
Nitrite	7.28	6.36
Bromide	8.66	8.16
Nitrate	9.06	8.94

[a] No peak observed.

Thus the concentration of anion A in an elution peak reduces the benzoate concentration by an equivalent amount. However, the eluent equilibrium is dynamic, and it is always 20 % ionized. This means that 80 % of the hydrogen and benzoate ions in the exchange reaction come from *molecular* benzoic acid and are converted into highly ionized H^+A^-. This effect enhances the detection sensitivity considerably because the H^+ counterion of the A^- has an unusually high equivalent conductance. A discussion of the phenomenon is given in Section 6.3.3.6.

Several other organic acids have been investigated as eluents in ion chromatography [14]. One objective was to find an eluent acid that is not adsorbed by the resin matrix and that attains equilibrium faster than benzoic acid.

Relative retention times of the sample anions in Table 6.8 show rather small differences from one acid eluent to another in a few cases. However, iodide has relative t' that would make separations easier with some eluents than with others.

Although only monovalent anions are included in Table 6.8, similar effects are observed with some divalent anions. Salicylic acid will elute sulfate in 5.2 min and thiosulfate in 7.3 min under the conditions in Table 6.7. However, the weaker eluents such as succinic or nicotinic acid require an excessive amount of time to elute divalent anions.

Table 6.9 lists the relative retention time of the chloride sample anion with the acid eluents chosen for this study. The adjusted retention time of chloride decreases as the retention time for the eluent anion increases and it also decreases as the amount of ionization of the eluent increases. The most satisfactory separations were obtained with either nicotinic acid or succinic acid as the eluent acid.

Table 6.8. Relative adjusted retention times (t'/t' Cl) of various anions with different acid eluents (Cl = 100)[a]. (From Ref. [14]).

Anion	Succinic	Nicotinic	Benzoic	Salicylic	Fumaric	Citric
HPO_4^{2-}	0.73	0.87	0.81	0.82	0.67	0.68
IO_3^-	0.77	0.76	0.91	0.80	0.88	0.79
NO_2^-	0.91	1.00	0.89	0.79	0.87	1.04
Cl^-	1.00	1.00	1.00	1.00	1.00	1.00
BrO_3^-	1.14	1.11	1.05	1.00	1.29	1.28
$MeSO_3^-$	1.22	1.11	1.00	1.03	1.19	1.22
Br^-	1.41	1.42	1.21	1.03	2.27	2.08
$EtSO_3^-$	1.60	1.52	1.16	1.13	1.87	1.79
NO_3^-	1.62	1.57	1.29	1.13	1.70	1.76
ClO_3^-	2.31	2.14	1.56	1.29	2.65	2.45
$PrSO_3^-$	3.96	3.31	2.11	1.71	4.84	4.22
I^-	7.63	3.91	2.53	2.82	10.48	9.53

[a] Conditions: TMA–XAD–I, 0.027 mequiv/g, 30–37 m, 1 mM acid eluent.

Table 6.9. Relationship of chloride retention time and the affinity of the acid eluent for the resin.

Acid eluent	Chloride RT (min)	Eluent RT (min)	Eluent % dissociation
Nicotinic	25.70	3.9	11
Succinic	14.90	3.1	25
Benzoic	8.20	8.1	22
Citric	5.16	14.1	58
Fumaric	3.87	14.7	62
Salicylic	1.14	37.2	63

6.3.4 System Peaks

Eluent dips or system peaks and their causes were first described by Gjerde and Fritz [13]. Stevens et al. [3] described the effect of the system peak in suppressed ion chromatography. Called a carbonate dip, the system peak was said to be the absent peak (from the injection) of the carbonic acid that is retained by the unexhausted portion of suppressor column.

As an eluent is pumped through a column, the resin becomes equilibrated with the eluent. The desired process is an ion-exchange equilibrium in which the anions on the resin are displaced by the eluent anion. However, a second equilibrium process can occur in which the molecular form of the eluent is sorbed by the resin matrix. A system peak results from a change in this latter equilibrium which is caused by injection of a sample. If the sample pH is more basic than the eluent, then part of the sorbed eluent is ionized and desorbed. The system peak which elutes later is from the readsorption of the eluent; hence, in this case it is a decreasing peak of dip. If the sample pH is higher than the eluent pH then some additional molecular eluent is sorbed on the resin. A positive peak results from the desorption of the "excess" eluent.

The retention time of the system peak is greater for resins of high surface area and porosity, and when the polarity of the resin matrix and eluent are more alike. In general, system peak effects are lowered by choosing a more polar eluent or resin, or by raising the eluent pH.

6.3.5 Scope of Anion Separations

Anions of strong acids may be separated in acidic solution (as in Fig. 6.10) or at a basic pH. Weak acid anions require a basic solution to exist in the anionic form. Separation of borate, silicate, sulfide, cyanide as well as the anions of two stronger acids is shown in Fig. 6.11. An alkaline solution of sodium benzoate was used as the eluent. Carbonate may also be separated under alkaline conditions. Separation of these anions by suppressed IC is usually not attempted because they are converted to the non-conducting molecular form by the acidic suppressor.

Although a very large number of organic anions may be separated by ion chromatography, non-suppressed conductance is not always a viable detection method. Anions of small organic acids can usually be detected but the conductivity of more

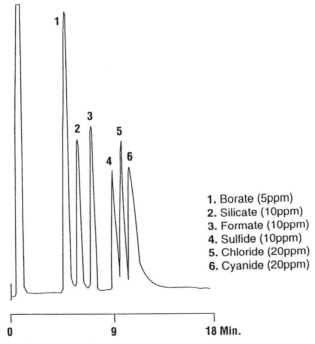

1. Borate (5ppm)
2. Silicate (10ppm)
3. Formate (10ppm)
4. Sulfide (10ppm)
5. Chloride (20ppm)
6. Cyanide (20ppm)

Figure 6.11. Separation of weak acid anions. Conditions: Alltech Anion/R column, sodium hydroxide/sodium benzoate eluent, 1.5 mL/min, conductivity detection. Hydroxide is the primary eluent for this separation and the weak acid anions are detected with indirect conductivity. A small amount of benzoate is added to speed the elution of the anions (Courtesy Alltech).

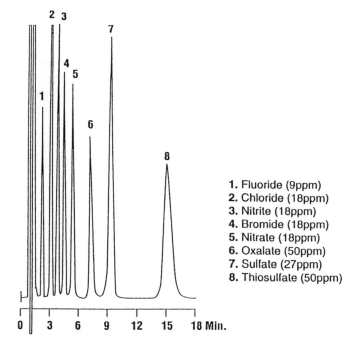

Figure 6.12. Common anions, nonsuppressed conductivity detection. Conditions: Hamilton PRP-X110 column, 2 mM *p*-hydroxybenzoic acid/0.1 mM sodium thiocyanate, pH 9.30, 2.0 mL/min (Courtesy Hamilton Co.).

bulk organic anions tends to approach that of the eluent anion, thus giving poor detection sensitivity. If needed, a certain percentage of an organic solvent (methanol, acetonitrile, etc.) may be added to the aqueous eluent to improve the analyte solubility.

6.3.6 Sensitivity

The term "sensitivity" will be used here in its technically correct context as the change in detector signal per unit concentration. It should not be confused with detection limits that are dependent on baseline noise. (Baseline noise is dependent on the signal magnitude, temperature variations, the electronics, etc. and varies from instrument to instrument.) In this section we shall examine the factors that determine the sensitivity that can be attained in anion chromatography.

The approach taken utilizes Table 4.1 of limiting equivalent conductances of various anions and cations to estimate the changes in conductance that will occur using various chromatographic techniques. The calculations are verified by comparing the experimental and calculated background conductances of several eluents. Expected changes in conductances are calculated for separation of selected anions using different chromatographic techniques.

Experimental measurements of the conductance of various eluents were made by Gjerde and Fritz [13].The eluent was pumped through a Wescan model 213 conductivity detector until a steady reading was obtained. Most of the measurements were made at 22.0–22.3 °C, although a few were at a slightly higher or lower temperature.

The cell constant was measured at 22.3 °C with a 1.00×10^{-3} solution of potassium chloride. It was found to be 33.0 cm^{-1} as calculated from the equation: $k = GK$, where k is the known specific conductance of the potassium chloride in µS, G is the conductance in µS, and K is the cell constant.

Next, the conductance G was measured for several different eluents. For comparison, the conductances of the same eluents were calculated from a table of limiting ionic equivalent conductances, with the equation:

$$G_{(\mu S)} = \frac{(\lambda_+ + \lambda_-)CI\,10^6}{10^3\,K} \tag{6.1}$$

where λ is the limiting equivalent conductance of the cation or anion, C is the normality, and I is the fraction of the eluent which is ionized. The factor 10^6 converts G into µS. The term is divided into 10^3 because there are (approximately) 10^3 cm^3 in 1 liter. The following illustrates this calculation for 100 % ionized, 2×10^{-4} M potassium benzoate:

$$G = \frac{(7.35 + 32.4)(2 \times 10^{-4})10^6}{1000(33.0)} = 0.64 \; \mu S \tag{6.2}$$

The calculated and experimental values in Table 6.10 show quite reasonable agreement. The accuracy of these comparisons is limited by use of *limiting* equivalent conductances instead of the equivalent conductances at the concentrations actually used, imprecise temperature control, and somewhat limited accuracy in measuring the cell constant. Nevertheless, calculations from a table of equivalent conductances provide a useful estimation of expected experimental results.

The results in Table 6.10 show that it is better to use a sodium salt than a potassium salt as the eluent, because of the lower equivalent conductance of the sodium ion. A lithium salt would, presumably, be even better. The background conductance of some of the benzoate or phthalate eluents used in single-column chromatography are not greatly different from the background conductance of the more concentrated carbonate-bicarbonate eluent used in suppressed anion chromatography. For example, the background conductance of 2.0×10^{-4} M sodium benzoate is actually lower than with the carbonate-bicarbonate eluent that is widely used in suppressed ion chromatography. Thus, the baseline noise can be similar in the two chromatographic methods. However, higher eluent concentrations, up to 5 mM are more common in non-suppressed work.

Table 6.10. Background conductances of eluents[a]

Eluent, concentration	Temp. [°C]	Conductance [μS]	
		Measured	Calculated
NaBz, 2×10^{-4} M, pH 7.0	22.0	0.50	0.50
NaBz, through catex	21.3	0.92	0.98
KBz, 2×10^{-4} M, pH 7.0	22.0	0.65	0.64
HBz, 8.4×10^{-4} M,	22.2	2.15	2.32
NaHCO₃, 2×10^{-4} M, pH 7.5	22.8	0.53	0.57
NaHCO₃, through catex	22.7	0.15	
Pure H_2O	23.0	0.042	
Pure H_2O, through catex	23.0	0.050	
NaHCO₃, 0.003 M + Na₂CO₃, 0.0024 M	22.8	20.9	
NaHCO₃ + Na₂HCO₃, through catex	22.8	0.63	
KPh, 2×10^{-4} M, pH 6.7	22.1	1.36	
KPh, through catex	22.0	1.93	
K₃Cit, 2×10^{-4} M, pH 7.6	22.0	2.37	
K₃Cit, through catex	22.0	1.96	

[a] Bz = benzoate; pH = phthalate; Cit = citrate; catex = cation exchange column in H^+ form.

6.3.6.1 Conductance of a Sample Peak

The general equation for detector response in ion chromatography was derived in Chapter 5. The following is a more specific derivation that takes into account that the eluent and sample are not always completely ionized.

In ion chromatography, once the pseudo peak has passed through the column and detector the total cation and anion concentration in solution is constant and equal to that of the eluent. Since we are dealing with an ion-exchange process, a sample anion moves down the column by exchange with an equivalent amount of eluent anion. When a sample anion passes through the detector, the eluent anion concentration is decreased by an amount equivalent to the concentration of the sample anion. Thus, the change in conductance will be determined by the relative equivalent conductances of the sample and eluent conductance (or at least different) than the eluent anion to give a positive peak.

The relative sensitivity of different methods of anion chromatography can be compared by calculating the change in conductance resulting from the replacement of the eluent anion with an equivalent concentration of a sample anion. The eluent cation also affects the conductance, but in single-column chromatography the cation contribution to conductance remains constant. When a suppressor is used and the eluent cation is exchanged for a new cation, the cation contribution will of course change. Therefore, we shall consider only the cations which are present in the detector cell so that the equations developed can apply to both single-column and suppressor ion chromatography.

Equation 6.1 can be used to calculate the change in conductance when a sample anion is present. Let C_E be the concentration of the eluent anion, and C_S be the concentration of sample anion.

Taking all forms of eluent and sample into account:

Eluent: $HE \rightleftharpoons H^+ + E^-$ $\qquad\qquad\qquad\qquad\qquad\qquad$ (6.3)

Sample: $HS \rightleftharpoons H^+ + S^-$ $\qquad\qquad\qquad\qquad\qquad\qquad$ (6.4)

And the ion exchange reaction is:

$$\text{Resin-S} + \text{H}^+ + \text{E}^- \; \rightleftharpoons \; \text{Resin-E} + \text{H}^+ + \text{S}^-$$
$$\qquad\qquad\quad \updownarrow \qquad\qquad\qquad\qquad\qquad \updownarrow \qquad\qquad\qquad (6.5)$$
$$\qquad\qquad\quad \text{HE} \qquad\qquad\qquad\qquad\qquad \text{HS}$$

The background conductance (taken from Eq. 6.1), where there is no sample present, is given by the equation:

$$G_B = \frac{(\lambda_{E+} + \lambda_{E-})C_E \mathbf{I_E}}{10^{-3}\,K} \qquad\qquad\qquad\qquad (6.6)$$

where E^+ and E^- are the eluent cation and anion, respectively, and $\mathbf{I_E}$ is the fraction of the eluent that is ionized:

$$\mathbf{I_E} = \frac{[E^-]}{[HE] + [E^-]} \qquad\qquad\qquad\qquad\qquad (6.7)$$

and $C_E = [HE] + [E^-]$ $\qquad\qquad\qquad\qquad\qquad\qquad$ (6.8)

The concentration of E^- during a sample peak elution will be $(C_E - C_S)\mathbf{I_E}$. The concentration of S^- during a sample peak elution is $C_S \mathbf{I_S}$. The conductance of the eluent and sample during a peak elution, G_S, is described with Eq. 6.9. The conductance is the sum of the contributions from the eluent ions and the sample ions.

$$G_S = \frac{(\lambda_{E+} + \lambda_{E-})(C_E - C_S)\mathbf{I_E}}{10^{-3}\,K} \qquad\qquad\qquad (6.9)$$

Subtraction of Eq. 6.8, the background conductance, from Eq. 6.9, the peak conductance, gives the *change* in conductance when a sample peak is eluted:

$$G_S - G_B = \Delta G = \frac{(\lambda_{E+} + \lambda_{S-})\mathbf{I_S} - (\lambda_{E+} + \lambda_{E-})\mathbf{I_E}}{10^{-3}\,K}(C_S) \qquad (6.10)$$

where ΔG is the detector response.

This equation can be used for organic acid eluents as well as for sodium or potassium salts or acids. It shows that sensitivity is dependent on the extent of ionization of an acidic eluent as well as the difference in equivalent conductances between the sample and eluent ions. A sample peak is approximately Gaussian and the sample anion concentration C_S changes accordingly as the elution peak develops.

Equation 6.10 represents the relationship between detector response and sample concentration under all conditions of ion chromatography with conductivity detection. K and λ are constant under most conditions in ion chromatography. I_E and I_S are dependent on the ionization of the respective acid and on the eluent concentration and pH. The eluent is a buffer and I_E and I_S are relatively constant unless the buffer capacity is exceeded.

Calculation of the expected conductance change with a benzoic acid eluent is accomplished with Eq. 6.10. An 8.40×10^{-4} M benzoic acid eluent is calculated to be 23.8 % ionized and thus contains 2.00×10^{-4} M benzoate ion. Experimentally it is similar in eluting power to 2.00×10^{-4} M sodium benzoate. If the ionization of benzoic acid is assumed to remain constant, it can be shown that molecular benzoic acid, as well as the benzoate ion, will exchange with a sample anion, A^-, to give H^+A^-. If the latter is completely ionized, the change in conductance, ΔG, will be:

$$\Delta G = [(\lambda_{H^+} + \lambda_{A^-}) - (\lambda_{H^+} + \lambda_{HBz^-})\, 0.238]\, [A^-] \tag{6.11}$$

Equation 6.11 can be simplified under conditions where the eluent and sample are completely ionized. If it is assumed that I_S and I_E are 1, then from Eq. 6.10:

$$\Delta G = \frac{(\lambda_{S^-} - \lambda_{E^-})}{10^{-3}\, K}\, (C_S) \tag{6.12}$$

Thus, it is seen that in single-column chromatography, ΔG (μS) is proportional to the difference in equivalent conductances of the sample and eluent anions:

$$\Delta G \, \alpha \, (\lambda_{S^-} - \lambda_{E^-})\, C_S \tag{6.13}$$

In Table 6.11, Eq. 6.9 is used to estimate the change in conductance for various eluents. The calculations are made for an anion with an equivalent conductance of 70. For comparison the background conductance of the eluent is also given at the eluent concentration typically used.

With carboxylic acid eluents, the sensitivity increases as the eluent dissociation decreases, but decreased dissociation also decreases the eluting power of the eluent. There is a trade-off between sensitivity and elution power with this system. The easiest way around this problem is to increase the eluent concentration. The sensitivity will change slightly as the extent of dissociation changes, but the strength of the eluent can be greatly increased.

The pH of the eluent acid should also be expected to affect the detection sensitivity for various anions. For anions of strong acids (chloride, bromide, nitrate, etc.) the detection sensitivity should be excellent because the sample anion and the highly mobile hydrogen counterion are completely ionized. However, with anions of weaker acids (acetate, fluoride, etc.), some of the anion will be present as the molecular acid and the detection sensitivity will consequently be less. Furthermore, the detection sensitivity of weak acid anions will become progressively less favorable as the pH of the eluent acid becomes more acidic.

Table 6.11. Estimated change in conductance with various eluents [a]

Eluent Concentration	Background G[b] [μmhos]	ΔG [μmhos]
NaBz, 2.00×10^{-4} M, pH 7.0	0.50	1.27×10^3 (X)
NaBz through catex	0.92	6.94×10^3 (X)
HBz, 8.40×10^{-4} M	2.15	9.96×10^3(X)
K_2Ph, 2.00×10^{-4} M, pH 7.0	1.36	0.94×10^3 (X)
K_3Cit, $2.00x~10^{-4}$ M, pH 7.0	2.37	0.61×10^3 (X)
NaOH 1.25×10^{-3} M		3.88×10^3 (X) nonsuppressed
		12.7×10^3 (X) after suppression
NaHCO$_3$, 0.003 M + Na$_2$CO$_3$, 0.024 M	0.63	12.7×10^3 (X)

[a] The calculations are made for a sample anion having an equivalent conductance of 70; X represents the normality of the sample anion.

[b] $K = 33.0$ cm^{-1}.

The data in Table 6.11 indicate that suppressed conductivity methods will give better sensitivity and lower detection limits than non-suppressed. In fact, use of a hydroxide eluent will give even better sensitivity than the carbonate-bicarbonate eluent listed in the table. Figure 6.12 shows a separation of 1 ppm each of seven common anions with non-suppressed conductivity detection. A separation on the same column with a different eluent and suppressed conductivity detection shows significantly higher peaks for the same anion concentration (Fig. 6.13).

6.3.7 Limits of Detection

Detection limits depend on several factors in addition to the conductivity system used. These include the sample volume, the goodness of temperature control and the inherent sensitivity of the conductivity detector used. With a direct injection of a 50 μL sample, the detection limits for chloride, nitrate and sulfate in drinking water have been estimated to be around 10 ppb using suppressed conductivity [15]. With the same volume of sample the detection limits using non-suppressed conductivity are probably around 100 to 200 ppb. However, considerably lower limits of detection are possible with non-suppressed conductivity of a carboxylic acid eluent is used. Table 6.12 gives detection limits for a 100 μL water sample.

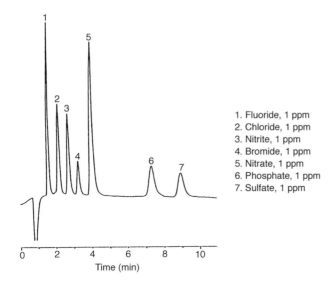

1. Fluoride, 1 ppm
2. Chloride, 1 ppm
3. Nitrite, 1 ppm
4. Bromide, 1 ppm
5. Nitrate, 1 ppm
6. Phosphate, 1 ppm
7. Sulfate, 1 ppm

Figure 6.13. Common anions, suppressed conductivity detection. Conditions: Hamilton PRP-X110S, 150 × 4.1 mm, 1.7 mM sodium bicarbonate/1.8 mM sodium carbonate/0.1 mM sodium thiocyante, 2.0 mL/min (Courtesy Hamilton Co.).

Table 6.12. Limits of detection (ppb).[a] (From Ref. [14]).

Ion	Succinic acid	Nicotinic acid
Chloride	26	5
Fluoride	20	4
Nitrite	35	9

[a] Eluents were 1 mM at natural pH, 500 × 2 mm glass column, 0.027 mequiv/g XAD-1, trimethylamine exchanger

6.4 Optical Absorbance Detection

6.4.1 Introduction

Ion chromatography was originated with conductivity detection and has grown up with suppressed and non-suppressed conductivity as the leading form of detection. However, conductivity detection does have some drawbacks. Temperature affects conductance, so a constant-temperature device is an essential part of the IC system. Suppressed conductivity requires an extra suppressor unit in the system that inevitably contributes to peak broadening. Conductivity is in effect a "universal" detection system because all ions passing through the detector contribute to the measured conductance. Although universal detection can be an advantage, it is often better to use a more selective detector. Several such detectors will be discussed in the following sections, but a UV-Vis spectrophotometric detector is one of the very best.

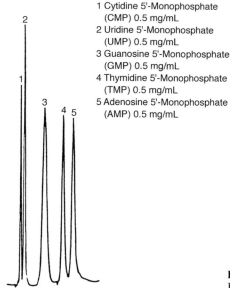

1 Cytidine 5'-Monophosphate
(CMP) 0.5 mg/mL
2 Uridine 5'-Monophosphate
(UMP) 0.5 mg/mL
3 Guanosine 5'-Monophosphate
(GMP) 0.5 mg/mL
4 Thymidine 5'-Monophosphate
(TMP) 0.5 mg/mL
5 Adenosine 5'-Monophosphate
(AMP) 0.5 mg/mL

Figure 6.14. Separation of nucleotides. Conditions: Hamilton PRP-X100 column 150×4.1 mm; A: 25 mM citric acid pH 5.4; B: 17:3 25 mM citric acid:acetonitrile, linear gradient 0 to 50 % B from 0 to 10 min (Courtesy Hamilton Co)

It is probably best to use a variable-wavelength UV-Vis spectrophotometer and not a fixed wavelength instrument. These detectors are very stable and have excellent sensitivity. We have obtained an excellent baseline and chromatograms when the detector was set on 0.003 AU full scale. The variable-wavelength feature is useful because detectors of this type can be used to take advantage of rather small differences in absorbance spectra of various ions.

Absorbance detection has been applied to ion analysis through two different approaches: direct detection of the sample ion and indirect detection. In some cases, a post-column, color-forming reagent can be added to the column eluate to detect sample ions.

6.4.2 Trace Anions in Samples Containing High Levels of Chloride or Sulfate

The determination of trace levels of ions in the presence of very high levels of sample matrix ions is a fairly common analytical problem. Seawater, for example, contains approximately 0.50 M sodium chloride, corresponding to a chloride concentration of approximately 18 000 ppm. Ion-chromatographic separation of trace anions becomes difficult in this situation because the large chloride peak may obscure peaks of the analyte anions. Even when a selective detection method is employed where the matrix anion gives little or no signal, it can still exert a major influence on the final chromatogram by causing variable retention times and a loss of chromatographic efficiency.

A good answer to this problem is to use an eluent containing the same ion as the sample matrix. Marheni, Haddad and McTaggart used eluents containing chloride or sulfate for samples containing a high concentration of chloride or sulfate, respectively [16].

With a 50 × 4.6 mm column packed with a polymethacrylate anion exchanger of 30 ± 3 μ/mL capacity, the chromatographic behavior of several anions was studied by direct photometric detection at 210 nm. Linear plots were obtained for log k (log retention factor) vs. log NaCl concentration in the eluent (Fig. 6.15).

The eluent chosen for practical separations contained 15 mM sodium chloride and 5 mM phosphate buffer at pH 6.5. Separation of the ten anions gave very similar chro-

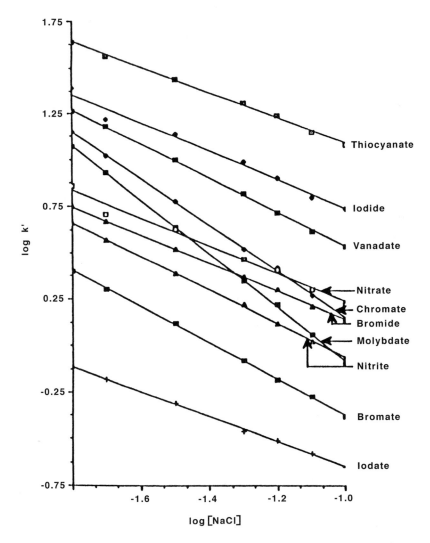

Figure 6.15. Variation of log k with the concentration of chloride in the eluent. (From Ref. 16] with permission).

matograms for samples in water alone and in 20 000 ppm chloride, although the retention times were slightly shorter in the latter case.

The mechanism of this separation is as follows: Introduction of the sample containing 20 000 ppm chloride (560 mM) results in a solution at the top of the column that is much higher than the eluent (15 mM chloride). But because the anion exchanger is already in the chloride form, the sample zone of very high chloride can pass rapidly through the column at the same linear flow rate as the eluent. Although the retention factors of the analyte anions are lower in the presence of the high sample chloride concentration, their k values quickly return to normal as soon as the high-chloride zone has passed.

It should be noted that this method succeeds in part because chloride has little effect on the detection of the trace ions at 210 nm. Detection limits were slightly higher in a high-salt sample than when the sample was in water (Table 6.14). Sample peak heights were only slightly lower in the samples containing a high chloride concentration. For example the peak height of Br^- in 20 000 ppm chloride was 85 % of that in water; the value for nitrate was 95 % of that in water.

6.4.3 Direct UV Absorption

Table 6.13 lists some common inorganic and organic anions that absorb sufficiently at wavelengths slightly above 210 nm to be detected by direct UV. Many anions (fluoride, chloride, perchlorate, sulfate and others) are UV-transparent at these wavelengths. All that is needed is to select a suitable non-absorbing anion for use in the mobile phase and carry out the IC separation using direct UV detection of the analyte ions. For example, sulfate is a good eluting ion by virtue of its 2– charge. Perchlorate is an even stronger eluting ion because of its strong affinity for anion exchange sites on the column packing. Direct UV detection is probably the easiest and one of the most sensitive ways to monitor the separation of UV-absorbing anions.

Both inorganic and organic anions may be determined by anion chromatography with direct UV detection. Figure 6.14 shows a separation of nucleotides with 25 mM citric acid buffered at pH 5.4 as the eluent. A gradient elution with increasing amounts of acetonitrile was used to speed up elution of the later peaks.

Table 6.13. Anions that can be detected by direct UV (see also Chapter 6).

Inorganic Anions	Organic Anions
Azide	Acetate
Bromate	Aromatic carboxylates and sulfonates
Bromide	Citrate
Chromate	Formate
Dichromate	Glutamate
Iodide	Lactate
Iodate	Malate
Nitrate	Malonate
Nitrate	Oxalate
Noble metal complexes	Phenolates
($AuCl_4^-$, $PtCl_6^{2-}$, $PdCl_4^{2-}$, etc.)	Succinate
Perrhenate	Tartrate
Thiosulfate	
Vanadate	

Table 6.14. Detection limits (ppm) with a 10-µl injection volume. Data from Ref. [16].

Anion	Eluent 15 mM NaCl	
	Sample in H_2O	Sample in 20 000 ppm Cl^-
IO_3^-	0.12	0.26
BrO_3^-	0.32	1.2
NO_3	0.03	0.06
Br^-	0.16	0.29
NO_3^-	0.08	0.12
MoO_4^{2-}	0.18	0.40
CrO_4^{2-}	0.70	1 .32
VO_3^-	1.70	2.6
I^-	0.57	0.82
SCN^-	0.60	1.0

6.4.4 Indirect Absorbance

In this detection mode, an eluent anion is chosen that absorbs in the visible or UV spectral region. The elution of sample anions is monitored by measuring the decrease in absorbance at the detection wavelength as transparent sample ions replace a fraction of the absorbing eluent anions. Under conditions where the solute anion (S^-) is fully dissociated, which is the usual case for IC, the change in absorbance (ΔA) is given by:

$$\Delta A = (\varepsilon_{S^-} - \varepsilon_{E^-})\, C_{S^-}\, (b) \qquad\qquad (6.14)$$

where ε_{S-} and ε_{E-} are the molar absorptivities of the solute and eluent anions, C_S is the molar concentration of the solute anion, and b is the cell pathlength. Ordinarily, the eluent anion and detection wavelength will be selected so that the absorbance will decrease and the solute peaks will be in the negative direction.

The concentration of the eluent anion must be high enough to elute the sample ions within a reasonable time and to exceed the concentration of any sample anion at its peak maximum. However, if the concentration of E^- greatly exceeds that of S^-, the background absorbance will be relatively high and the noise of peak detection will be poor. So while a reasonable eluent concentration is needed to promote ion exchange with the column, the concentration should still be as low as possible to reduce noise. To achieve the best sensitivity, ε_{E-} should be as high as possible so that the difference in molar absorptivities in Eq. 6.14 will be large.

In summary, the eluting anion selected for indirect UV detection should have a strong affinity for the ion exchanger so that a relatively low concentration can be used, and it should have a high molar absorptivity.

Several anions have proven to be effective for indirect photometric detection. Figure 6.16 shows an excellent separation of common anions with 2.0 mM potassium phthalate, pH 6.0, with indirect detection at 280 nm. By switching to 4.0 mM *p*-hydroxybenzoic acid, pH 8.5 with 2.5 % methanol, detection at 310 nm, and reducing the column length to 10 cm, the same anions plus phosphate could be separated in only 2.3 min [17].

A sodium molybdate eluent with indirect detection at 250 nm was found to provide an excellent separation and a very sensitive detection of inorganic anions. The separa-

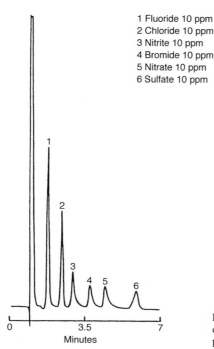

1 Fluoride 10 ppm
2 Chloride 10 ppm
3 Nitrite 10 ppm
4 Bromide 10 ppm
5 Nitrate 10 ppm
6 Sulfate 10 ppm

Figure 6.16. Separation of common anions with indirect UV detection at 280 nm. Eluent: 2.0 mM potassium hydrogen phthalate pH 6.0, 1.2 mL/min (Courtesy Hamilton Co).

tion in Fig. 6.17 was performed on a latex-coated anion exchanger at 27 μequiv/g capacity [18].

Miura and Fritz investigated the use of polycarboxylic acid salts as eluents in anion chromatography [19]. At alkaline pH values, 1,3,5-benzenetricarboxylic acid (BTA)

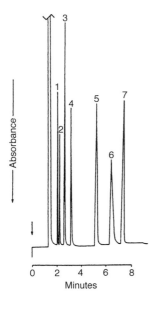

Figure 6.17. Separation of 7 common anions on a latex-coated column with a sodium molybdate eluent run at 0.75 mL/min. Indirect spectrophotometric detection was used at 250 nm with 0.05 a.u.f.s. Peaks: 1 = ethylsulfonate; 2 = propylsulfonate; 3 = chloride; 4 = nitrite; 5 = bromide; 6 = nitrate; 7 = sulfate (10–20 ppm each anion). (From Ref. [18] with permission).

exists as the 3– anion and pyromellitic acid (1,2,4–5-benzenetetracarboxylic acid) can exist as the 4– anion. Because of their high charge, these ions are very effective as eluents. Eluents containing 1.2 mM phthalate 2–, 0.2 mM BTA 3–, and 0.1 mM pyromellitate 4– were approximately equivalent in their ability to elute common inorganic anions. Indirect spectrophotometric detection of sample anions was always much better than conductivity detection. In practice, however, it is more difficult in indirect UV to establish the optimum conditions for separation and detection. Figure 6.18 shows a separation of 2.0 ppm concentrations of anions using very dilute (0.02 mM) pyromellitate eluent at pH 7.0 with indirect detection at 220 nm.

6.5 Potentiometric Detection

Potentiometric detection of anions is feasible when an electrode is available that responds quickly, reversibly and reproducibly to the concentration (or more precisely to the activity) of sample ions. It is often possible to detect a given ion or class of ions with excellent selectivity. For example, solid-state or crystalline ion selective electrodes have been used in IC to detect halide anions. The fluoride ion-selective electrode is particularly selective [20,21]. A copper wire electrode has been used to detect anions such as iodate, bromide and oxalate [22].

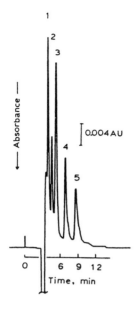

Figure 6.18. Separation of anions with 0.02 mM pyromellitate eluent (pH 7.0), detection at 220 nm, 25-cm column. Peaks: 1 = Cl⁻ (1.25 ppm); 2 = NO₂⁻ (2 ppm); 3 = SO₄²⁻ (2 ppm); 4 = Br⁻ (2 ppm); 5 = NO₃⁻ (2 ppm). (From Ref. [19] with permission).

A metallic silver electrode responds rapidly and reproducibly to the activity of free silver ions in solution. At 25 °C:

$$E = E° + 0.05915 \log a_{Ag+} \tag{6.15}$$

where E is the electrode potential (V) and $E°$ is the standard reduction potential for $Ag^+ + e^- = Ag°$. If the silver metal is coated with a slightly soluble solid (AgX), a_{Ag+} and hence E is determined by the concentration of X^- in solution via the solubility product of AgX.

$$K_s = a_{Ag+} \times a_{X-}; \, a_{Ag+} = K_s/a_{X-} \tag{6.16}$$

Combining Eqs. 6.15 and 6.16:

$$E = E° + 0.05915 \, (\log K_s - \log a_{X-}) \tag{6.17}$$

Lockridge et al. [23] prepared a number of silver wire electrodes coated with one of the following insoluble silver salts: AgCl, AgBr, AgI, AgSCN, Ag₂S and Ag₂PO₄. The coating process was performed by anodic oxidization of a silver wire for 3 to 5 min in a solution of the appropriate anion. The electrodes coated with AgCl or AgSCN were found to give the best response with either of these electrodes. A reproducible potentiometric response could be obtained for any of the halide or pseudohalide anions but only if the electrode was conditioned by dipping it into a solution of the analyte 3 or 4 times before use as a detector. Electron micrograph photos showed the surface to be a composite of several silver salts covering the underlying silver chloride precipitate on the electrode surface.

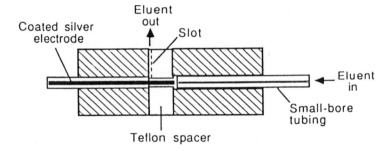

Figure 6.19. Flow cell for potentiometric detection with coated silver electrode. (From Ref. [12] with permission).

The cell for potentiometric detection is shown in Fig. 6.19. The eluate from the IC column flows past the coated silver indicator electrode and then out past a small silver, silver chloride reference electrode (not shown in the figure). The eluate itself serves as a salt bridge between the two electrodes.

A silver wire electrode coated with AgCl has excellent selectivity. An ion chromatogram obtained with 4.5 mM sodium perchlorate as the eluent gave sharp peaks for 1 mM chloride, bromide, iodide, thiocyanate and thiosulfate but no response to equi-

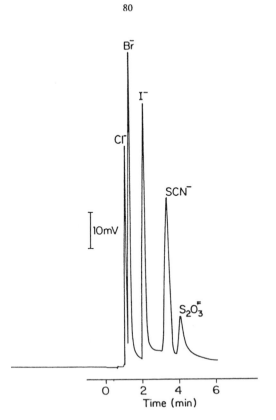

Figure 6.20. Gradient elution with potentiometric detection. Eluent: 3.5–10.0 mM sodium perchlorate; flow-rate, 1.6 mL/min; injection volume, 20 μL; analyte concentration, 1.0 mM. (From Ref. [23] with permission).

molar concentrations of nitrate, phosphate, carbonate, sulfate and acetate. The background response of this potentiometric system is virtually unaffected by changes in the concentration of sodium perchlorate or sodium sulfate eluent. This permits the use of gradient elution as shown in Fig. 6.20 where a gradient of 3.5 to 10.0 mM sodium perchlorate was used to obtain a fast separation of five halide and pseudohalide ions.

6.6 Pulsed Amperometric Detector (PAD)

The principles of the PAD were discussed in Chapter 4. Perhaps the major use of the PAD in anion chromatrography has been for the detection of carbohydrates. At pH around 11 or higher sugars become anionic and can be separated by anion chromatography. For many years a somewhat awkward post-column derivatization reaction was used for detection of carbohydrates after a chromatographic separation, but the use of a PAD now provides a simple and direct detection method.

All carbohydrates (aldoses and ketoses) and polyalcohols produce a large anodic peak response at *ca.* +0.15 V. The peak for glucose corresponds to a reaction with n approaching 10 equiv/mol for fluid velocities typical of flow-through detection cells. This n value is consistent with an oxidative cleavage for the C_1–C_2 and C_5–C_6 bonds to form two moles of formate and one mole of dicarboxylate dianion [24].

Either a gold or platinum electrode may be used in a PAD, although gold electrodes are more popular. One reason is that dissolved oxygen contributes a cathodic response at a platinum electrode over the entire useful range for anodic detection. Serious oxygen interference at the gold electrode can be avoided by careful selection of detection potential.

A simple but useful example of pulsed-amperometric detection is shown in Fig. 6.21 where glucose, fructose and a trace of sucrose are determined in honey by anion chromatography. Much more complex samples can be resolved using gradient elution. This is demonstrated in Fig. 6.22 where 18 carbohydrates were separated. Elution of the later peaks is speeded up by gradually reducing the eluent pH to inhibit ionization of the carbohydrates. However, post-column addition of 0.4 M sodium hydroxide was needed to restore the effluent to a pH sufficiently alkaline for effective pulsed amperometric detection.

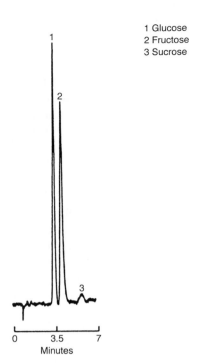

1 Glucose
2 Fructose
3 Sucrose

Figure 6.21. Sugars in honey. Conditions: Hamilton RCX-10 column 250 × 4.1 mm, 30 mM sodium hydroxide, pulsed amperometric detection with dual gold electrode. E_1 = 350 mV for 166 ms, E_2 = 900 mV for 166 ms, E_3 = –800 mV for 333 ms (Courtesy Hamilton Co).

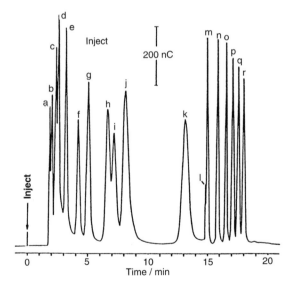

Figure 6.22. Separation of carbohydrate mixture. Pulsed amperometric detection (model I) at Au; Ag/AgCl reference; E_{det} = +0.10 V (t_{det} = 610 ms, t_{del} = 400 ms, t_{int} = 200 ms); E_{oxd} = +0.80 V (t_{oxd} = 120 ms); E_{red} = –0.60 V (t_{red} = 200 ms). Column: Dionex AS-6 Carbopac. Solvents: (A) 100 mM NaOH, (B) 50 mM NaOH + 0.5 M NaOAc, (C) H_2O. Elution: isocratic (0–6 min) with A–C (50:50); linear gradient (6–15 min) to A–B (50:50); isocratic (15–21 min) with A–B (50:50). Post-column addition: 0.4 M NaOH. Peaks: a = inositol; b = xylitol; c = sorbitol; d = mannitol; 3 = fucose; f = rhamnose; g = arabinose; h = glucose; i = xylose; j = fructose; k = sucrose; l = unknown; m = maltose; n = maltotriose; o = maltotetraose; p = maltopentaose; q = malto-hexaose; r = maltoheptaose. (From Ref. [26] with permission).

6.7 Inductively Coupled Plasma Atomic Emission Spectroscopy (ICP-AES)

ICP-AES is often used to determine the concentrations of various elements in a sample. However, an element may be present in a variety of chemical forms or species. By coupling an ICP-AES detector to an ion-chromatographic column, a more complete description of the sample species can be obtained. Such a coupling generally requires a nebulizer to introduce the column effluent into the ICP. Conventional pneumatic nebulizers operate at about 1 mL/min sample flow and may introduce as little as 1 % of the sample into the plasma. A newer direct-injection nebulizer (DIN) operates at sample flow rates only 5 to 10 % that of a conventional nebulizer [25].

The separation of various arsenic species is a good example of the application of ICP-AES detection to anion chromatography [26]. A microbore column 10 cm × 1.7 mm I.D. was used with a low flow rate (<100 μL/min). The column was packed with a low-capacity anion-exchange material (0.05 mequiv/g and solution containing 5 mM ammonium carbonate and 5 mM ammonium bicarbonate at pH 8.6 served as the mobile phase. The column hardware was connected directly to the inlet of the DIN-ICP-AES via a short length of 0.3 mm I.D. PEEK tubing.

A separation of arsenite, arsenate and monomethylarsonate (MMA) is shown in Fig. 6.23. The detection limit for arsenic was 10 μg/L and the minimum detectable quantity was 100 pg. It was also possible to separate and detect selenium(IV) and (VI) under similar conditions.

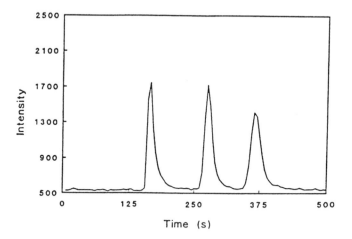

Figure 6.23. Anion-exchange separation of arsenic species. A 5.0 mM ammonium carbonate/5.0 mM ammonium bicarbonate eluent was used at 80 μL/min flow rate and a 10 μL sample volume. Peaks (in order); As(III), MMA, As(V) (from Ref. [27] with permission).

6.8 Inductively Coupled Plasma Mass Spectrometry (ICP-MS)

It is also possible to use mass spectroscopy (MS) detection for anionic species separated by IC. Limits of detection are very low and the use of MS permits the measurement of different isotopes of the same element.

The separation of selenite (SeO_3^{2-}) and selenate (SeO_4^{2-}) by anion chromatography with ICP-MS detection is a typical example [27]. The connection of the anion exchange column and direct injection nebulizer to the mass spectrometer and chromatographic conditions are similar to those employed with AES detection [26]. It is important to minimize the dead volume between the column and DIN, otherwise significant broadening of the separated peaks can occur.

Retention time was 3.4 min for SeO_3^{2-} and 9.0 for SeO_4^{2-}. Separation of 5 ng of each selenium species gave a RSD of 3.0 %. The absolute detection limit was approximately 15 pg for each species. The relative amounts of ^{78}Se and ^{74}Se could also be measured.

It is also possible to couple a small anion-exchange column to the MS with an electrospray for sample introduction. The conditions here are mild enough for the electrospray molecules to survive.

References

[1]   H. Small, T. S. Stevens and W. C. Bauman, Novel ion exchange chromatographic method using conductimetric detection, *Anal. Chem.*, 47, 1801, 1975.

[2]   S. Rabian, J. Stillian, V. Barreto, K. Friedman and M. Toofan, New membrane-based electrolytic suppressor device for suppressed conductivity detection in ion chromatography, *J. Chromatogr.*, 640, 97, 1993.

[3]   T. S. Stevens, J. C. Davis and H. Small, Hollow fiber ion exchange suppressor for ion chromatography, *Anal. Chem.*, 53, 1488, 1981.

[4]   J. Stillian, An improved suppressor for ion chromatography, *LC Mag.*, 3, 802, 1985.

[5]   D. L. Strong and P. K. Dasgupta, Electrodialytic membrane suppressor for ion chromatography, *Anal. Chem.*, 61, 939, 1989.

[6]   A. Henshall, S. Rabin, J. Statler and J. Stillian, Recent development in ion chromatography detection: the self-regenerating suppressor, *Am. Lab.*, 24, 20R, 1992.

[7]   D. T. Gjerde and J. V. Benson, Suspension postcolumn reaction detector for liquid chromatography, *Anal. Chem.*, 62, 612, 1990.

[8]   P. E. Jackson, P. Jandik, J. Li, J. Krol, G. Bondoux and D. T. Gjerde, Practical applications of solid-phase reagent conductivity detection in ion chromatography, *J. Chromatogr.*, 546, 189, 1991.

[9]   D. T. Gjerde, D. J. Cox, P. Jandik and J. B. Li, Determination of analytes at extreme concentration ratios by gradient ion chromatography with solid-phase reaction detection, *J. Chromatogr.*, 546, 151, 1991.

[10]   D. T. Gjerde and J. S. Fritz, Effect of capacity on the behavior of anion-exchange resins, *J. Chromatogr.*, 176, 199, 1979.

[11]   D. T. Gjerde, J. S. Fritz and G. Schmuckler, Anion chromatography with low-conductivity eluents, *J. Chromatogr.*, 186, 509, 1979.

[12]   D. T. Gjerde, G. Schmuckler and J. S. Fritz, Anion chromatography with low-conductivity eluents. II, *J. Chromatogr.*, 187, 35, 1980.

[13]   D. T. Gjerde and J. S. Fritz, Behavior of various benzoate eluents for ion chromatography, *Anal. Chem.*, 53, 2324, 1981.

[14]   J. S. Fritz, D. L. DuVal and R. E. Barron, Organic acids as eluents for single-column anion chromatography, *Anal. Chem.*, 56, 1177, 1984.

[15]   J. Weiss, Ion Chromatography, VCH, Weinheim, Germany, 2nd Ed. 1995, p348.

[16] Marheni, P. R. Haddad and A. R. McTaggart, On-column matrix elimination of high levels of chloride and sulfate in non-suppressed ion chromatography, *J. Chromatogr.*, 546, 221, 1991.

[17] Hamilton Co., Reno, NV, USA. Hamilton applications handbook, Application #94.

[18] L. M. Warth, J. S. Fritz, J. O. Naples, Preparation and use of latex-coated resins for anion chromatography, *J. Chromatogr.*, 462, 165, 1989.

[19] Y. Miura and J. S. Fritz, Benzenepolycarboxylic acid salts as eluents in anion chromatography, *J. Chromatogr.*, 482, 155, 1989.

[20] M. P. Keuken, J. Slanina, P.A.C. Jongjan and F. P. Bakker, Optimization of ion chromatography using commercially available detection system and software, *J. Chromatogr.*, 439, 13, 1988.

[21] R. T. Talasek, Determination of fluoride in semiconductor process chemicals by ion chromatography with ion-selective electrodes, *J. Chromatogr.*, 20, 179, 1985.

[22] P. W. Alexander, P. R. Haddad and M. Trojanowicz, Potentiometric detection in ion chromatography using a metallic copper indicator electrode, *Chromatographia*, 20, 179, 1985.

[23] J. E. Lockridge, N. E. Fortier, G. Schmuckler and J. S. Fritz, Potentiometric detection of halides and pseudohalides in anion chromatography, *Anal. Chim. Acta*, 192, 41, 1987.

[24] D. C. Johnson, D. Dobberpuhl, R. Roberts and P. Vandeberg, Pulsed amperometric detection of carbohydrates, amines and sulfur species in ion chromatography – the current state of research, *J. Chromatogr.*, 640, 79, 1993.

[25] D. R. Wiederin, R. E. Smyczek and R. S. Houk, On-line standard additions with direct nebulization for inductively coupled plasma mass spectrometry, *Anal. Chem.*, 63, 1626, 1991.

[26] D. T. Gjerde, D. R. Wiederin, F. G. Smith and B. W. Mattson, Metal speciation by means of microbore columns with direct-injection nebulization by inductively coupled plasma atomic emission spectroscopy, *J. Chromatogr.*, 640, 73, 1993.

[27] S. C. K. Shum and R. S. Houk, Elemental speciation by anion exchange and size exclusion chromatography with detection by inductively coupled plasma mass spectrometry with direct injection nebulization, *Anal. Chem.*, 65, 2972, 1993.

7 Cation Chromatography

7.1 Separation Principles and Columns

The basic principles of cation chromatography are very simple. A suitable cation-exchange column is used in conjunction with the auxiliary equipment typical of liquid chromatography: eluent reservoir, pump, injection loop, guard column, detector and detector cell, and a data acquisition device. After equilibration of the system so that a steady baseline is obtained, an eluent is pumped through the system, a sample is injected and the cationic analytes are separated. A common separation mechanism is one in which the sample cations are pushed at different rates down the column by the cations in the mobile phase. However, a number of cations have very similar selectivity coefficients for the cation exchanger and cannot be separated by this method. A second general separation method uses a complexing reagent (or mixture of reagents) to move the sample cations down the column by partial complexation. This latter method has been used very successfully for separation of metal cations such as the divalent transition metals and the lanthanides.

Several types of cation-exchange columns have been used successfully for cation chromatography. Almost all of the cation exchangers are of low exchange capacity so that an eluent of a fairly low ionic concentration can be used. Low-capacity cation-exchange resins are obtained by superficial sulfonation of styrene-divinylbenzene copolymer beads as originally described by Small et al. [1] and by Fritz et al. [2]. The resin beads are treated with concentrated sulfuric acid and a thin layer of sulfonic acid groups is formed on the surface. The final capacity of the resin is related to the thickness of the layer and is dependent on the type of resin, the bead diameter, and the temperature and time of contact with the sulfuric acid. Typical capacities range from 0.005 to 0.1 mequiv/g compared to 5 mequiv/g for conventional cation-exchange resins.

It can be easily appreciated that, compared to a conventional cation-exchange resin, the diffusion path length is reduced because the unreacted, hydrophobic resin core restricts analyte cations to the resin surface. This results in faster mass transfer of the cations and consequently in improved separations. Also, because of the rigidity of the resin core, there is less tendency for the bead to compress. This means that higher flow rates (at relatively low back pressures) can be used than would be possible with conventional resins. Superficially functionalized resins are stable over the pH range of

1 to 14 and swelling problems are minimal. The selectivity of the superficial cation-exchange resins for ions is similar to that observed for conventional resins.

Over the years the performance of cation exchange materials for ion chromatography has improved significantly. The major types of columns commercially available are listed in Table 7.1.

Table 7.1. Cation-exchange columns.

Functional group	Type	Examples
Sulfonic acid	polymeric, surface sulfonated	Hamilton PRP-x200, Wescan Cation R
Sulfonic acid	polymeric, latex coated	Dionex CS3, CS10, CS11
Carboxyl	polymeric	Cetac ICSep CN2
Carboxyl	polymeric	Dionex CS12, CS14
Carboxyl	silica, polybutadiene–maleic acid coated	Alltech universal cation
Mixed	polymeric pellicular	Dionex CS5A
Mixed	carboxyl and phosphonate	Dionex CS12A
Mixed	carboxyl, phosphonate, crown ether	Dionex CS15
Mixed	silica base; carboxyl and crown ether	Alltech Cation HC

The surface-sulfonated cation-exchange columns from Wescan are packed with 10 μm spherical resins with an exchange capacity of 0.1 mequiv/g. The Hamilton PRP × 200 resin has a similar capacity (0.035 mequiv/g) and is available in 3 μm , 10 μm and 12–20 μm particle size.

Dionex Corp. offers several cation exchangers with a sulfonic acid function using latex technology similar to that of latex anion exchangers. A weakly sulfonated polystyrene–divinylbenzene substrate 10 μm in diameter is covered with a surface layer of fully aminated latex beads about 50 nm in diameter. These latex particles are agglomerated on the substrate surface by both electrostatic and van-der-Waals interactions. Then the substrate is covered by a second layer of latex beads containing a sulfonic acid to provide the actual cation exchange function of the doubly-coated substrate. In the CS3 columns the sulfonated latex beads have a diameter of about 250 nm and a cross-linking of 5 %. Even with the double coating, these cation-exchange materials are stable in actual use.

The Dionex CS10 materials are prepared in a similar manner to the CS3 except that the first layer of aminated latex particles is covalently bound. This imparts better mechanical and chemical stability. The high degree of cross-linking of the microporous support makes the CS10 compatible with solvents such as methanol and acetonitrile. The CS11 column listed in Table 7.1 is similar in selectivity to the CS10 but offers a 1.75-fold increase in capacity.

The ICSep CN2 Cation Exchange (Cetac Technologies, division of Transgenomic, Inc.) columns employ porous PS/DVB polymers with sulfonic and carboxylic acid functional groups to provide separation of cations. The ICSep columns are compatible with a wide variety of eluents including strong acids, organic acids and chelators. The column provides separation capabilities for Group I and II metals and ammonium species using either suppressed conductivity or UV detection modes. However, it is most useful for transition metal separations.

More recently, the trend in cation chromatography has been to resins with a weaker acid function. A carboxylic acid functional group will be completely in the anionic form at pH values somewhat higher than its pK_a value. By operating at more acidic pH values, ionization of the carboxyl group is diminished and the ion-exchange affinity of the resin is consequently decreased. Thus, control of eluent pH as well as the ionic concentration of the eluent can be used to obtain a desired separation. The Alltech "universal cation" column contains silica spheres coated with a butadiene–maleic acid copolymer to provide the ion-exchange function. Maleic acid has two carboxyl groups per molecule, $pK_1 = 2.0$, $pK_2 = 6.3$ and therefore retains some anionic character at more acidic pH values than a simple carboxylic acid.

Two ion exchange resins from Dionex contain carboxyl groups. The CS14 column is recommended for gradient separation of amine cations.

Several cation-exchange columns with mixed functional groups are listed in Table 7.1. A phosphonic acid group has a pK_a value intermediate between the weakly acidic carboxylic acid (pK_a 4 to 5) and the strongly acidic sulfonic acid group. Incorporation of both carboxyl and phosphonate groups in the same resin enables one to gradually reduce the effective ion exchange capacity over a larger pH range. Starting about pH 5, gradual reduction of the eluent pH first reduces the anionic carboxyl function by protonation but the phosphonate remains completely in the anionic form. Further decreases in pH convert more of the phosphonate to the protonated form, resulting in further decreases in effective ion-exchange capacity.

At least two sources offer resins containing a crown ether function in addition to carboxyl or carboxyl and phosphonate (see Table 7.1). In particular, the crown ether function strengthens the affinity of the ion exchanger for potassium(I) by virtue of complexation. These resins are recommended for determination of trace sodium and ammonium in various treated water- or wastewater samples.

7.2 Separation with Ionic Eluents

In this type of separation the analyte cations compete with the eluent cation for ion-exchange sites and move down the column at different rates. The ionic eluent selected depends on the cations to be separated, the type of separation column and on the detector. In many cases an aqueous solution of a strong acid such as hydrochloric, sulfuric or methanesulfonic acid is a satisfactory eluent. Sample cations commonly separated include the following: alkali metal ions (Li, Na^+, K^+, Rb^+, Cs^+), ammonium, magnesium, alkaline earths (Ca^{2+}, Sr^{2+}, Ba^{2+}), and various organic amine and alkanolamine cations. Most other metal cations are separated with a weakly complexing eluent.

7.2.1 Suppressed Conductivity Detection

With suppressed conductivity detection, an acidic cationic eluent is used to separate the sample cations. The column effluent with zones of separated cations passes directly into the suppressor unit containing an anion-exchange membrane in the hydroxide form. The eluent cation is neutralized and the counteranions associated with the sample metal ions are exchanged for the more highly conducting hydroxide ion.

For example, if a dilute nitric acid eluent is used and sodium and potassium sample ions are to be separated, the following reactions take place in the suppressor unit:

(1) Eluent: $H^+NO_3^- + Res–OH^- \rightarrow ResNO_3^- + H_2O$

(2) Sample: $Na^+NO_3^-, K^+NO_3^- + Res\text{-}OH^- \rightarrow Na^+OH^-, K^+OH^- + Res\text{-}NO_3^-$

The background conductivity is very low after the eluent passes through the suppressor unit; theoretically it is that of pure water. The equivalent conductance of sample ions is high; it is the sum of conductances of the alkali metal cation and the hydroxide counter ion.

Modern suppressors for cation chromatography are both efficient and self-regenerating. The principles are similar to the suppressors for anion chromatography, described in Chapter 6. The mechanism of suppression for a cation self-regenerating suppressor is illustrated in Fig. 7.1 and described in some detail by Rabin et al. [3]. Suppressors for cation chromatography are limited to those cations that do not form precipitates with the hydroxide ions from the suppressor.

Excellent separations of all the alkali metal cations plus ammonium in 10 min or less with a strong acid (sulfonic acid) cation exchanger and a dilute solution of a strong acid as the eluent (incomplete sentence). However, divalent metal cations are more strongly retained by this column and require either an eluent containing a divalent cation or a more concentrated solution of the H^+ eluent.

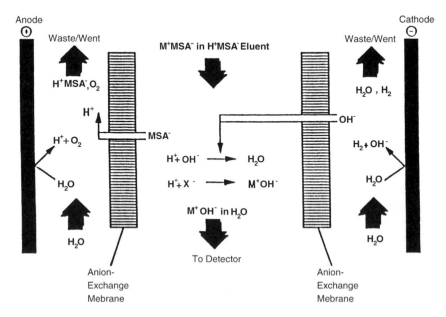

Figure 7.1. Mechanism of suppression for the cation self-regenerating suppressor. H^+MSA^- = Methanesulfonic acid. From Ref. [3] with permission.

By using an ion exchanger with carboxyl groups or with both carboxyl and phosphonate groups, it is possible to separate both monovalent alkali metal cations and certain divalent metal cations in a single run. A dilute solution of a strong acid such as methanesulfonic acid is generally used as the eluent. Particular emphasis has been placed on separation of Li^+, Na^+, NH_4^+, K^+, Mg^{2+} and Ca^{2+} because these ions are found in many types of samples. Fig. 7.2 shows a separation of all six metal ions at concentrations of 0.5 to 5.0 ppm on a Dionex CS12A bifunctional column [4]. Dilute sulfuric acid was used as the eluent with suppressed conductivity detection.

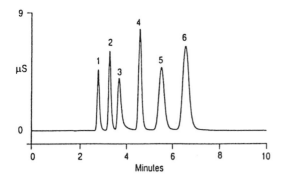

Figure 7.2. Fast separation of alkali and alkaline earth metals and ammonium with CS 12A. Eluent: 15.5 mM sulfuric acid. Detection: suppressed conductivity. Peaks: 1 = lithium (0.05 mg/L); 2 = sodium (2 mg/L); 3 = ammonium (2.5 mg/L); 4 = potassium (5 mg/L); 5 = magnesium (2.5 mg/L); 6 = calcium (5 mg/L). From Ref. [4] with permission.

Gradient elution is feasible with modern suppressed conductivity detection. Several aliphatic amines were separated as the protonated amine cations in Fig. 7.3 using a sulfuric acid gradient [5]. The increasing acidity served to reduce the effective exchange capacity of the ion exchanger and thereby speed up elution of the larger monoamines and the diamines.

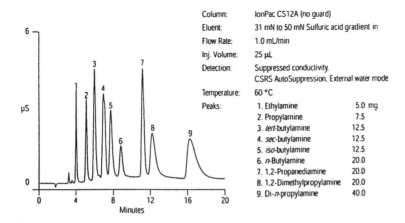

Figure 7.3. Gradient separation of aliphatic amines with a sulfuric acid gradient (Courtesy Dionex Corp).

A different type of gradient was used to separate the metal ions and quaternary ammonium cations in Fig. 7.4. In this case an eluent of fixed acidity was used (11 mM sulfuric acid) but the gradient increased the acetonitrile content of the mobile phase from 10 % to 80 % over 15 min [6]. The purpose of the acetonitrile gradient was to decrease the hydrophobic affinity of the higher amine salts for the ion exchanger.

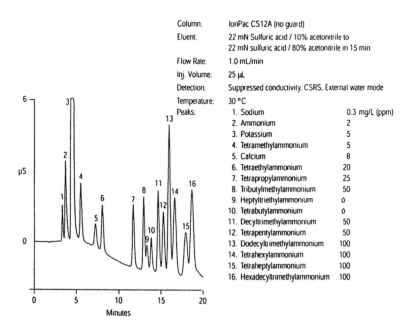

Column:	IonPac CS12A (no guard)	
Eluent:	22 mN Sulfuric acid / 10% acetonitrile to	
	22 mN sulfuric acid / 80% acetonitrile in 15 min	
Flow Rate:	1.0 mL/min	
Inj. Volume:	25 μL	
Detection:	Suppressed conductivity, CSRS, External water mode	
Temperature:	30 °C	
Peaks:	1. Sodium	0.3 mg/L (ppm)
	2. Ammonium	2
	3. Potassium	5
	4. Tetramethylammonium	5
	5. Calcium	8
	6. Tetraethylammonium	20
	7. Tetrapropylammonium	25
	8. Tributylmethylammonium	50
	9. Heptyltriethylammonium	6
	10. Tetrabutylammonium	6
	11. Decyltrimethylammonium	50
	12. Tetrapentylammonium	50
	13. Dodecyltrimethylammonium	100
	14. Tetrahexylammonium	100
	15. Tetraheptylammonium	100
	16. Hexadecyltrimethylammonium	100

Figure 7.4. Gradient elution of hydrophobic quaternary ammonium ions. (Courtesy Dionex Corp).

7.2.2 Non-Suppressed Conductivity Detection

With modern columns and dilute solutions of a strong acid as the eluent, cations may be separated and detected with excellent sensitivity by direct conductivity as well as by suppressed conductivity [2]. The basis for direct conductivity detection is that the highly conductive H^+ (equivalent conductance = 350 S cm^2 equiv^{-1}) in the eluent is partially replaced by a cation of lower conductance when a sample zone passes through the detector. For example, the equivalent conductance of Li^+, Na^+, and K^+ is 39, 50 and 74 S cm^2 equiv^{-1}, respectively. The decrease in conductance on an equivalent basis can be calculated as follows: Background: $H^+ + NO_3^- = 350 + 71 = 421$ (S cm^2 equiv^{-1}). Sample peaks: $Li^+ + NO_3^- = 39 + 71 = 110$, a decrease of 311; $Na^+ + NO_3^- = 50 + 71 = 121$, a decrease of 300; $K^+ + NO_3^- = 74 + 71 = 145$, a decrease of 276.

Riviello et al. made a careful comparison of conductivity changes in cation chromatography between direct- and suppressed conductivity detection [7]. The calculation example is outlined in Fig. 7.5. The change in conductivity, ΔG, is actually slightly greater with non-suppressed conductivity. However, the noise is much higher in the non-suppressed detection mode. Noise may be defined as the random signal that

Example of Direct Conductivity Detection	**Example of Suppressed Conductivity Detection**

Example of Direct
Conductivity Detection

Example of Suppressed
Conductivity Detection

Eluent: 1 mM HNO_3
Sample: 0.1 mM $NaNO_3$

$G_E = (\lambda_{H^+} + \lambda_{NO_3^-}) C_E = (350 + 71)1$
$\quad = 421\ \mu S$

$G_S = (\lambda_{H^+} + \lambda_{NO_3^-}) C_E + (\lambda_{Na^+} + \lambda_{NO_3^-}) C_S$
$\quad = (350 + 71)0.9 + (50 + 71)0.1$
$\quad = 391\ \mu S$

$\Delta G_{S\text{-}E} = 391 - 421 = -30\ \mu S$

∴ A decrease in conductivity is observed when the sample cation elutes.

Eluent: 1 mM HNO_3
Sample: 0.1 mM $NaNO_3$

$G_E \cong 0$

$G_S = (\lambda_{H^+} + \lambda_{NO_3^-}) C_E + (\lambda_{Na^+} + \lambda_{OH^-}) C_S$
$\quad = (350 + 71)0 + (50 + 198)0.1$
$\quad = 25\ \mu S$

$\Delta G_{S\text{-}E} = 25 - 0 = 25\ \mu S$

An increase in conductivity is observed when the sample cation elutes.

C_E = eluent concentration (mM)

C_S = sample concentration (mM)

λ_X = equivalent conductance of X (μS/mM)

G_E = background conductance of eluent,
$\quad G_E = (\lambda_{H^+} + \lambda_{X^-}) C_E$

G_S = conductance in the sample band,
$\quad G_S = (\lambda_{H^+} + \lambda_{X^-}) C_E + (\lambda_{C^+} + \lambda_{X^-}) C_S$

$\Delta G_{S\text{-}E}$ = sample signal

Figure 7.5. Comparison of conductivity changes with direct and suppressed conductivity detection. Adapted from Ref. [7].

results from chemical background (conductance in this case) temperature fluctuations, hydraulics and electronics. It was found that noise is proportional to background conductivity. Pump noise and detector/electronic noise also increase with increased background conductivity. Temperature control is critical for direct detection and slightly improves suppressed detection.

A major difference between the two detection modes is that detection limits for alkali metal ions were 12–21 times lower with suppressed detection [7]. Nevertheless, cations may be separated and quantified with direct detection down to fairly low concentrations. Another advantage is that the eluent remains acidic and metal does not precipitate as easily as it might in the high pH environment of the suppressor. Figure 7.6 shows a separation of several metal cations including cesium and strontium at concentrations averaging only 1.0 ppm.

Ion exchangers with both carboxyl and crown ether functions have different selectivities than simple carboxyl materials. In particular, potassium is complexed by the

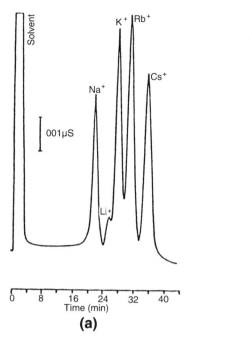

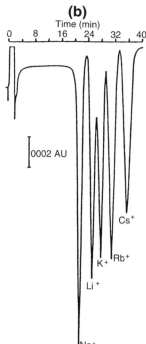

Figure 7.8. Chromatogram obtained with 0.2 mM 2,6-dimethylpyridine at pH 6.35 as eluent by use of (a) direct conductivity and (b) indirect UV absorption detection. Sample: 15 μL of a solution containing 2×10^{-5} M of each of the indicated ions. From Ref. [8] with permission.

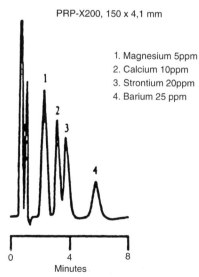

PRP-X200, 150 x 4,1 mm

1. Magnesium 5ppm
2. Calcium 10ppm
3. Strontium 20ppm
4. Barium 25 ppm

Conditons: 3,5 mM Cupric Sulfate. Isocratic. Ambient. 2mL/min. Indirect UV 220 nm.

Figure 7.9. Separation of metal ions on a PRP-X200 column, 150 × 4.1 mm. Eluent 3.5 mM cupric sulfate, indirect UV detection at 220 nm (Courtesy Hamilton Co).

between Na^+ and K^+ with an acidic eluent such as dilute nitric acid, Li^+ always elutes before Na^+.

The separations in Fig. 7.8 were performed with 15 μL of a sample containing 2×10^{-5} M of each sample ion. This corresponds to an absolute amount of only 0.30 nmol of each ion. The detection limits of several inorganic ions have been calculated for aromatic bases as eluent components using a 100 μL sample. The results given in Table 7.3 show detection limits in the low ppb concentration range. Most of the detection limits were lower for indirect UV detection than for direct conductivity.

There are certainly many other possibilities for direct spectrophotometric detection. A separation of several divalent metal ions is shown in Fig. 7.9 with 3.5 mM cupric sulfate as the eluent and indirect detection at 220 nm.

Table 7.3. Detection limits (ppb) for typical eluents, calculated for a 100 μL injection. Indirect UV detection.

Eluent	Li^+	Na^+	K^+	NH_4^+	Mg^{2+}	Ca^{2+}	Sr^{2+}
2-Phenylethylamine	1	2	2	3	1	4	12
Benzylamine	1	0.4	0.4	0.5	11	63	290
4-Methylbenzylamine	0.2	1	1	1	5	16	47

7.3 Effect of Organic Solvents

7.3.1 Separation of Amine Cations

As in anion chromatography, the IC separation of organic cations has long been known, or at least suspected, that its mechanism involved more than simple ion exchange. Hoffman and co-workers [9,10] have shown that two mechanisms occur in such cases: ion exchange and hydrophobic interaction between the sample cations and the resin matrix. For example, these authors showed that the slopes of the linear plot of log k' vs. carbon number for protonated amine cations decrease going from 30 % acetonitrile (70 % water) to 70 % acetonitrile in the eluent. This is due to lower hydrophobic interaction in the 70 % acetonitrile.

Dumont, Fritz and Schmidt studied cation chromatography in organic solvents containing little if any water [11]. Under these conditions solvation of the lipophilic part of the cation should be sufficient to virtually eliminate the hydrophobic interaction between the sample cations and the ion-exchange resin. In this way the true ion-exchange selectivity could be measured.

After trying several different inorganic acids, methanesulfonic acid was selected as the eluting acid for the separation of protonated amine cations. In IC of cations with H^+ as the eluting cation, k should vary according to the following equation:

$$\log k = -m \log H^+ + b$$

where *m* is the slope of a linear plot and *b* is a constant. Linear plots were obtained for the C_1–C_{10} *n*-alkylamines in methanol, ethanol, 2-propanol and acetonitrile. The slopes (*m*) were very close to the theoretical slope of –1 in the three alcohols and only a little less than –1 in acetonitrile.

The effect of solvent was studied by measuring the retention factor, *k*, for a series of protonated alkylamine cations with 25 mM methanesulfonic acid in the appropriate solvent as the eluent [11]. Ordinarily a plot of log *k* vs. the number of carbon atoms in such a homologous series would be linear. The slope of such a plot is at least in part an indication of the effect of the carbon chain on the retention factor. The retention factors were measured under identical conditions in each of four organic solvents. The *k* values of the alkylamines increased according to the solvent used in the following order: methanol, ethanol, 2-propanol, acetonitrile. However, in any given solvent the *k* values of the individual amines were almost constant from C1 to C10. Separation of the individual amines was not possible.

Two other types of organic cations did show enough difference in their retention factors for practical separations. Separation of aniline, *N*-methylaniline and *N,N*-dimethylaniline in 100 % methanol is shown in Fig. 7.10. Significant differences in *k* values of octylamine, dioctylamine and trioctylamine were observed in methanol, ethanol and 2-propanol.

These results suggest that a successful separation of organic cations by IC depends on differences in hydrophobic attraction between the solute ions and the ion exchanger as well as on differences in electrostatic attraction. Incorporation of an organic solvent in the eluent will increase the solubility of samples containing organic solutes. However, it is usually better to work with a mixed organic-aqueous eluent rather than

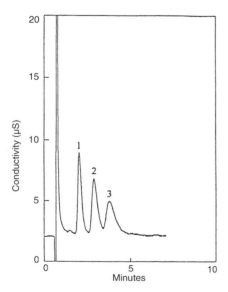

Figure 7.10. Separation of 12.5 ppm aniline (1), *N*-methylaniline (2) and *N,N*-dimethylaniline and (3) on a 5 cm sulfonated resin column (0.15 mmol/g). The eluent was 1 mM methanesulfonic acid in methanol at a flow-rate of 1 mL/min. From Ref. [1] with permission.

one that is entirely organic. Conductivity detection is feasible in organic-aqueous solutions or in 100 % organic of the lower alcohols.

7.3.2 Separation of Alkali Metal Ions

Ion chromatographic separations of the alkali metal cations are normally performed with sulfonated microporous polymeric resins or with resins coated with a sulfonated latex. Dumont and Fritz selected a lightly sulfonated macroporous resin with a high degree of cross linking for a study on alkali metal ion separations in organic solvents [12]. Such a resin would be less likely to undergo volume changes due to swelling and should be more compatible with organic solvents.

A separation of alkali-metal ions was first attempted in water alone using the lightly sulfonated macroporous cation exchanger with aqueous 3 mM methanesulfonic acid as the eluent. Under these conditions the sample cations exhibited very similar retention times.

When the macroporous resin column was used with the same acidic eluent in 100 % methanol, the chromatographic separation was improved considerably. Now the alkali-metal ions are solvated with methanol and the resin matrix is probably coated with a thin layer of methanol, which makes the ions and the resin surface more compatible with one another.

Although several factors may influence the selectivity of cation-exchange resins for 1+ metal cations, electrostatic attraction of the sulfonate groups within the ion-exchange resin for alkali-metal cations suggest that cations with the smallest ionic radii would be the most strongly retained. The Pauling radii in Table 7.4 would predict a chromatographic elution order of Cs^+, Rb^+, K^+, Na^+, Li^+, which is exactly the opposite of that observed in ion-exchange chromatography. However, hydrated ionic radii and approximate hydration number are in the opposite order to the Pauling radii, with Li^+ being the most highly hydrated. When separations are carried out in a nonaqueous solvent, the solvation of the alkali metal cations is apt to change. This could result in changes in elution characteristics of the ions.

Table 7.4. Ionic radii of alkali metal cations.

	Li^+	Na^+	K^+	Rb^+	Cs^+
Pauling radii (Å)	0.60	0.96	1.33	1.48	1.69
Hydrated radii (Å)	3.40	2.76	2.32	2.28	2.28
Approximate hydration number	25.3	16.6	10.5	10.0	9.90

The results of these studies [12] are summarized in Table 7.5. The use of non-aqueous solvents with macroporous cation-exchange resin permit several separations that are very difficult with aqueous eluents. Methanol was found to be the most favorable solvent due to the best combination of resolution and peak shape. Acetonitrile and ethanol, although producing broader peaks, are useful for separating ions that usually elute close together, Li^+/Na^+ and K^+/NH_4^+ respectively. Elution order in acetonitrile is reversed from that found with aqueous eluents: $Cs^+ < Rb^+ < K^+ < Na^+ < Li^+$. It was also discovered that addition of a crown ether, 18-crown-6, to the mobile phase improves both the peak shape and resolution of several ions.

Table 7.5. Capacity factors (k') in organic and mixed solvents with 0.5 mM methanesulfonic acid as the eluent.

Solvent:	Li^+	Na^+	K^+	Rb^+	Cs^+	NH_4^+
Water, 100 %	2.27	2.27	2.71	2.74	2.96	3.04
Methanol:						
25 %	2.66	2.47	2.77	2.80	2.98	3.13
50 %	3.10	3.07	3.31	3.39	3.80	3.70
75 %	4.43	5.07	6.32	7.30	8.33	5.82
100 %	2.08	2.82	3.73	4.33	5.15	3.09
Ethanol:						
25 %	2.75	2.57	2.78	2.78	2.96	3.13
50 %	3.25	3.09	3.37	3.48	3.76	3.78
75 %	4.61	5.14	6.90	7.62	8.76	5.98
100 %	1.86	3.84	7.24	8.84	9.87	2.12
2-Propanol:						
25 %	2.20	1.98	2.11	2.05	2.17	2.45
50 %	2.26	2.14	2.35	2.41	2.62	2.88
75 %	3.41	3.52	4.42	4.81	5.66	4.69
100 %	8.84	12.3	19.5	>20	>20	4.54
Acetonitrile:						
25 %	2.10	2.10	2.37	2.38	2.54	2.51
50 %	2.50	2.38	2.89	2.98	3.35	3.10
75 %	3.00	3.12	3.79	3.95	4.45	3.98
100 %	4.46	2.11	1.75	1.59	1.54	2.40

7.4. Separations with a Complexing Eluent

7.4.1 Principles

The separations discussed thus far have been possible because of differing affinities of the various cations for the cation-exchange column packing. However, the divalent transition metal ions, the lanthanides, and a number of others are very difficult to separate because their affinities for the ion exchanger are too similar. By adding a weak complexing ligand such as tartrate, citrate or oxalate to the eluent, the metal ions to be separated are partially complexed and converted to a non-charged or lower-charged metal complex. A certain fraction of each metal ion remains as the charged cation. The net effect is that the metal ions are eluted more rapidly. Owing to differences in the fraction that is not complexed, separation of the sample metal ions is enhanced.

The ligand selected and the concentration and pH of the eluent should be chosen so that complexation of the metal ions is only partial. If the metal ions are too strongly complexed, they will move too rapidly and no separation will occur.

For effective elution of divalent metal ions a divalent eluent cation such as the ethylenediammonium cation is also needed. Thus, divalent metal cations are eluted nicely by an eluent of 2.0×10^{-3} M ethylenediammonium tartrate, but not by an eluent containing sodium or ammonium tartrate.

In Chapter 5, the logarithm of the retention factor (log k) was shown to be a linear function of the logarithm of an ionic eluent (log E) as well as the logarithm of the exchange capacity (log C) (Eq. 5.19). When a complexing ligand is incorporated into the eluent, an additional term must be added to the equation:

$$\log k = \log \ \alpha_i - y/x \log [E] + y/x \log C + \text{Constant} \tag{7.1}$$

where α_i is the fraction of a metal ion that remains uncomplexed, y is the charge on the sample cation and x is the charge on the eluent cation. The value of α_i will depend on the formation constants of the metal-ligand complex, the concentration of ligand in the mobile phase and the pH.

As predicted by Eq. 7.1, a plot of log k or log adjusted retention time versus log α_i is linear for the trivalent lanthanides (Fig. 5.2) as well as for most of the divalent metal ions. By adjusting pH and the concentration of complexing ligand, log α_i can be varied and the retention time either increased or decreased.

Sevenich and Fritz found an eluent containing 2.0×10^{-3} M ethylenediammonium tartrate to be effective for separating several divalent metal cations [13]. Calculations from ionization constants showed ethylenediamine to be fully protonated (EnH_2^{2+}) at pH 5.0 or below. The adjusted retention times for several metal ions were obtained as a function of eluent pH (Table 7.6). The retention times increased in almost every case between pH 5 and 6. Furthermore, some minor extraneous peaks appeared in this pH region. A buffer pH of 4.5 was selected as giving generally the best results.

Table 7.6. Adjusted retention times (min) for elution with 2.0×10^{-3} M ethylenediammonium tartrate at varying pH [13].

pH	3.0	3.5	4.0	4.5	5.0	6.0
Metal ion			Adjusted retention time [min]			
Mg(II)	1.6	2.8	2.7	2.6	3.0	
Zn(II)	3.3	2.8	2.3	1.8	1.6	3.0
Co(II)	3.7	3.3	2.9	2.7	2.5	3.2
MN(II)	4.2	3.8	3.8	3.6	2.5	5.2
Cd(II)	5.2	4.7	4.7	4.0	4.2	5.5
Ca(II)	7.5	7.3	6.5	5.9	5.8	5.5
Sr(II)	11.8	11.8	10.8	10.1	9.8	10.7

Lead(II) shows a dramatic change in retention time as the pH is increased and complexation by the tartrate in the eluent becomes greater. For an example, t' for lead(II) is 18.3 min at pH 3.5, 10.0 min at pH 4.0, and 7.1 min at pH 4.5. A separation of 7 divalent metal ions with ethylenediammonium tartrate is shown in Fig. 7.11. Direct conductimetric was employed.

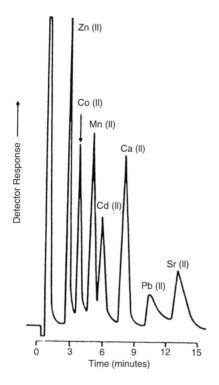

Figure 7.11. Separation of zinc(II) (10.3 ppm), cobalt(II) (9.1 ppm), manganese(II) (160.0 ppm), cadmium(II) (16.1 ppm), calcium(II) (17.1 ppm), lead(II), and strontium II (20.3 ppm). Eluent was 1.5 mM ethylenediammonium cation and 2.0 mM tartrate at pH 4.00. (Courtesy of G. J. Sevenich).

7.4.2 Use of Sample-Masking Reagents

Thus far, a weak complexing reagent has been added only to the eluent so that the sample metal ions will be only partially complexed. But what would happen if a weakly complexing eluent was used but a second strong but selective complexing ligand were added only to the sample? Under these conditions strongly complexed metal ions move rapidly through the column while the weakly complexed metal ions move more slowly and can be separated from one another [14]. If a selective complexing reagent could be employed, it should be possible to elute the strongly complexed metal ions quite rapidly and then to separate the remaining cations with the ethylenediammonium tartrate eluent described above. A further requirement would be that the auxiliary complexing agent must work in the 3 to 5 pH range needed for the eluent.

7.4.2.1 EDTA

Simple calculations showed that EDTA does not complex metal ions such as magnesium(II) and calcium(II) at pH 4 but it does complex many other metal cations. Therefore, experiments were performed in which ETDA is added to the metal ion sample and the column was eluted with ethlenediammonium tartrate as before [13]. The amount of EDTA used was more than enough to complex the metal ions present, but an unduly high concentration of EDTA was avoided. The results obtained show that conditions can easily be established whereby magnesium and the alkaline earth cation peaks are hardly affected but metal ions that form stable EDTA complexes at about pH 4 are rapidly eluted. Because EDTA is added only to the sample and not to the eluent, it moves rapidly through the column and appears as part of the pseudo peak.

Samples containing a large excess of iron(III) give extremely wide "pseudo peaks" when the ethylenediammonium tartrate is used. This excess of iron(III) will totally obscure the magnesium peak while calcium and strontium appear on the tail of the "pseudopeak".

Figure 7.12 shows a chromatogram of the same sample in which EDTA is added to complex the iron(III). The additional peak is from an iron(II) impurity in the iron(II) solution used. Work thus far indicated that any metal ion that has an EDTA formation constant of about 10^{15} or higher should be masked effectively by adding EDTA to the sample.

The very weak complexing of iron(II) by tartrate suggested that iron(II) might be determined quantitatively by cation chromatography. This was proved to be true by Fritz and Sevenich [14] who determined iron(II) in the presence of iron(III) and several other metal ions. Total iron in solution was determined after a preliminary reduction to iron(II) with ascorbic acid.

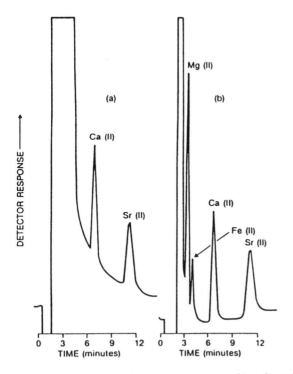

Figure 7.12. Separation of 0.10 mM each of Mg^{2+}, Ca^{2+}, Ca^{2+}, Sr^{2+} in the presence of a 100-fold excess of Fe^{3+} with (a) no EDTA in the sample (pH 1.7), and (b) 0.010 M EDTA added to the sample (pH 1.7). Eluent was 2.00 mM ethylenediammonium tartrate at pH 4.50. (Courtesy of G. J. Sevenich).

7.4.2.2 NTA as a Masking Reagent

Nitrilotriacetic acid (NTA) is essentially one half of an EDTA molecule and forms somewhat weaker complexes than EDTA. Calculations of α_M, the ratio of free metal ion to all forms of the metal in solution, showed that manganese(II), magnesium(II), and the alkaline earths are not complexed significantly by NTA below a pH of about 3.0. Iron(II) is not complexed at pH 2.0 or below, although some oxidation to iron (III)-NTA was found to occur. Many other metal ions form neutral anionic complexes with NTA at pH 3.0 and pass rapidly through the chromatographic column.

The ion-chromatographic determination of manganese(II) with NTA masking of other metal ions looked particularly promising. Addition of a constant excess of NTA to the sample and varying the pH showed an essentially constant peak height for manganese(II) between pH 2.0 and 3.0 [14]. For subsequent experiments, excess NTA was added to the sample and the pH was adjusted to 3.0.

Samples containing 50 µM each of magnesium(II), manganese(II), and calcium(II) plus an excess of another metal ion were treated with NTA and analyzed by ion chromatography using an ethylenediammonium tartrate eluent. No change in peak height was observed when a 10-fold or a 100-fold molar excess of iron(III), aluminum(III), copper(II), or thorium(IV) was added. A 10-fold excess of nickel(II) was also without effect, but a 100-fold excess did cause some change. Cobalt(II) and yttrium(III) are partly masked by the NTA. Lead(II), uranium(VI), cadmium(II), and zinc(II) are not masked. However, lead(II) elutes later than manganese, magnesium, and calcium and does not interfere.

Both peak heights and retention times are reproducible with excess NTA in the sample; the calibration curves are also linear. For example, a calibration curve for manganese(II), with a constant excess of aluminum(III) (5.0 mM) masked by 10 mM NTA was linear over a range of 0.020 to 0.50 mM manganese with a correlation coefficient of 0.999.

7.4.2.3 Sulfosalicylic Acid as a Masking Agent

The successful application of 5-sulfosalicylic acid in ion exchange separations was demonstrated by Fritz and Palmer [15]. At an appropriate pH, they found that aluminum(III), iron(III), uranium(VI), and vanadium(IV) are complexed and pass quickly through a cation exchange column. A number of other cations, including the rare earths, are not complexed and are retained on the column. Their application was simply a group separation, and no chromatographic separation of individual metal ions was involved.

Efficient and selective cation chromatographic separations are possible with an ethylenediammonium tartrate eluent at pH 4.5 and sulfosalicylic acid added to the sample before injection. At pH 4.5 zinc(II) can be chromatographed without interference from a 20-fold excess of thorium(IV), vanadium(IV), or uranium(VI), or from a 100-fold excess of iron(III) when sulfosalicylic acid is added to the sample [14].

Analysis of rare earth cations using sulfosalicylic acid to mask a large excess of aluminum(III) also proved quite successful. Linear calibration curves were obtained over at least 0.05 to 0.50 mM rare earth with 2.0 mM aluminum and 10 mM sulfosalicylic acid at pH 4.5.

7.4.3 Weak-Acid Ion Exchangers

Strong-acid (sulfonic acid) ion exchangers maintain their complete ionic capacity over a large pH range of 2–12. A weak-acid ion exchanger, such as one with carboxylic acid or phosphonic acid groups, loses its ionic capacity as the pH goes below the pK_a of the functional group. The eluent pH affects the separation of metal ions, thus adding an additional parameter for optimization.

The preparation and use of two new experimental resins was described by Morris and Fritz [16]. Both resins were prepared by simple Friedel-Crafts addition reactions. One has the carboxyl group attached to the benzene ring of a spherical polystyrene–divinylbenzene (PS-DVB) resin via a spacer arm of three carbon atoms. The carboxyl group is attached directly to the benzene ring in the other resin. Excellent ion chromatographic separations of metal ions are possible using these resins in conjunction with any of several complexing eluents.

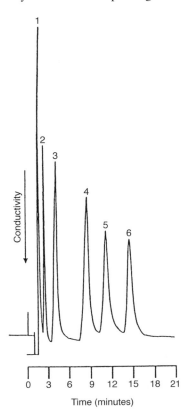

Figure 7.13. Chromatographic separation on resin I column (100 × 4.6 mm i.d.). Eluent conditions: 1.0 mM ethylenediammonium, 0.05 mM PDA (pH 5.4). Peaks: 1 = Zn^{2+} ; 2 = Na^+ ; 3 = Ca^{2+} ; 4 = Sr^{2+} ; 5 = Ba^{2+} ; 6 = Mg^{2+}. From Ref. [16] with permission.

A separation of several metal ions with an eluent containing 2,6-pyridinedicarboxylic acid (PDA) is shown in Fig. 7.13. The elution order is unusual. Zinc(II) is eluted even before Na$^+$ owing to its strong complexation by PDA (see Table 7.7). Magnesium(II) is only weakly complexed by PDA and elutes after Ca^{2+}, Sr^{2+} and Ba^{2+}. In other systems, Mg^{2+} almost always elutes before the alkaline earths.

Another example of a metal ion separation on a weak-acid column is shown in Fig. 7.14. Here, spectrophotometric detection at 530 nm was used after post-column addition of pyridylazo resorcinol (PAR) as a complexing reagent. At highly alkaline pH values forms colored complexes with metal ions that are more stable than the eluent complexing reagent (Chapter 4).

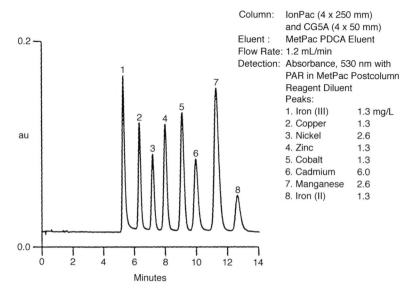

Column: IonPac (4 x 250 mm)
             and CG5A (4 x 50 mm)
Eluent :   MetPac PDCA Eluent
Flow Rate: 1.2 mL/min
Detection: Absorbance, 530 nm with
           PAR in MetPac Postcolumn
           Reagent Diluent
Peaks:
1. Iron (III)   1.3 mg/L
2. Copper   1.3
3. Nickel   2.6
4. Zinc   1.3
5. Cobalt   1.3
6. Cadmium   6.0
7. Manganese   2.6
8. Iron (II)   1.3

Figure 7.14. Separation of transition metals with the use of a complexing eluent containing PDA (Courtesy Dionex Corp).

Table 7.7. Logarithms of formation constants of selected metal complexes.

Reagent	Log formation constant								
	Cu^{2+}	Ni^{2+}	Co^{2+}	Zn^{2+}	Mn^{2+}	Mg	Ca^{2+}	Sr^{2+}	Ba^{2+}
Citric acid	5.60	5.11	4.83	4.70	3.70	3.25	3.18	2.81	2.55
PDA	8.80	6.60	6.35	6.43	4.70	2.02	4.30	3.50	3.13
Oxalic acid	4.53	3.70	3.25	3.43	2.60	2.10	1.66	1.25	1.02

7.5 Chelating Ion-Exchange Resins and Chelation Ion Chromatography

7.5.1 Fundamentals

The selectivity of ordinary cation-exchange resins for various metal ions is somewhat limited. However, if a suitable chelating functional group is built into a polymeric resin, it often is possible to take up only a small group of metal ions. Other chelating resins may complex a larger group of metal ions, but additional selectivity is attained through pH control. Chelating resins also are valuable in sorbing a desired metal ion (or small group of metal ions) from solutions containing a very high concentration of a non-complexed metal salt. Frequently the selectivity of a chelating resin is so great that a very short column can be used to retain the desired metals.

Preconcentration of selected metal ions is probably the main use of chelating resins in chemical analysis. Trace amounts of complexed metal ions may be concentrated from a large sample onto a very short column. Subsequent elution by acid breaks up the metal chelates and gives a much more concentrated solution of the metal ions for further analysis. However, a column packed with a chelating resin may also be used to separate sample metal ions based on differences in the strength of their chelates.

A few of the many types of chelating resins that have been synthesized are listed in Table 7.8. Resins containing the iminodiacetic acid (IDA) functional group have received particular attention. The material known as Chelex 100 has been available for many years but is not very efficient for chromatographic separation of metal ions. More modern IDA resins are more satisfactory. The IDA group forms chelates with a considerable number of metal ions and also provides good selectivity for metal ions that are not complexed. However, some problems can occur.

Table 7.8. Some typical chelating groups in resins.

Designation	Chelating group	Selective for:
IDA	$-N(CH_2CO_2H)_2$	Most M^{2+}, M^{3+}, M^{4+}
Amidoxime	$-C(=NOH)NH_2$	Divalent transition metals
DTC	$-NC(=S)S^-$	Ag^+, Hg^{2+}, Cu^{2+}, Cd^{2+}, Zn^{2+}, Co^{2+}, Ni^{2+}
Crown ether	$-OCH_2CH_2O-$ (cyclic)	K^+, some others
Hydroxamic acid	$-C(=O)(NOH)CH_3$	Fe^{3+}, Ti^{4+}, Th^{4+}, Zr^{4+}, Mo(VI) [19]

A high concentration of IDA groups on the chelating resin results in more complete complexation of metal ions from solution. A high concentration of complexing groups may also cause the resin to retain metal ions from a more acidic sample. However, stronger complexation of metal ions means that a more concentrated acid solution must be used to break up the complex and thereby desorb metal ions from the resin. The presence of excess acid may complicate the determination of sample ions for subsequent analysis by ion-chromatography or capillary electrophoresis.

An additional complication is that two kinds of metal ion uptake can occur with IDA resins. The desired kind of uptake involves chelation of metal ions with the nitrogen and carboxyl groups of the IDA as ligands. The other type is simple ion exchange of cations that are electrostatically attracted to the negatively charged carboxylate groups. This simple ion exchange can take up a significant amount of Na^+ or other unwanted cations, particularly if the resin contains a high concentration of IDA groups.

On balance, the best choice of an IDA resin might be one with a moderately low capacity (ca. 0.5 mequiv/g, for example) for retaining metal ions by chelation. A resin particle size of ca. 10 μm, should be used instead of the 40–50 μm size generally used in solid-phase extraction (SPE) cartridges.

For chromatographic separations it is important to use an efficient resin with the chelating groups readily accessible so that the chelation of metal ions is not sterically inhibited. The equilibrium between a divalent metal ion (M^{2+}) and a chelating resin (RL^-H^+) may be written:

$$M^{2+} + 2\ RL^-H^+ \rightleftharpoons (RL)_2M + 2\ H^+ \tag{7.2}$$

An acidic eluent is used to control this equilibrium so that the retention factor of the sample metal ion is in the desired range. Increasing H^+ concentration in the eluent weakens the chelates and speeds the elution. Separations of different metal ions will occur due to differences in the equilibrium constants for Eq. 7.2.

A second way to separate metal ions on a chelating resin column is to use a complexing eluent (E^-), such as oxalate tartrate, at a fixed pH. Here, a second equilibrium will come into play:

$$(RL)_2M + 2\ E^- \rightleftharpoons 2\ RL^- + ME_2 \tag{7.3}$$

Now the retention factor will be influenced by the type and concentration of L^- in the chelating resin, the pH of the eluent, and the type and concentration of E^- in the eluent.

Because formation and breakup of metal chelates is slower than a simple ion-exchange equilibrium, it is essential to select chelating resins with fast kinetics. When a complexing eluent is involved the kinetic situation may become more difficult. Now we can envision a still slower equilibrium between the metal chelate, $(RL)_2M$, and the eluent chelate, ME_2. As we shall see, this does not necessarily prevent effective separations with a complexing eluent, but it may still be an inhibiting factor.

7.5.2 Examples of Metal-Ion Separations

Bonn, Reiffenstuhl and Jandik were able to separate a number of divalent metal ions effectively in a single run using an IDA resin [17]. A silica based material (Nucleosil 300–7 of 7 μm diameter, 300 Å average pore size) was derivatized with γ-glycidoxypropyltrimethoxy silane, then iminodiacetic acid was covalently coupled to the epoxy activated surface. The final material was slurry packed into a 100 × 4.6 mm stainless steel column. A complexing eluent containing 10 mM citric acid plus

0.04 mM 2,6-pyridine dicarboxylic acid (PAD) gave a good separation of low ppm concentrations of Mg^{2+}, Fe^{2+}, Co^{2+}, Cd^{2+} and Zn^{2+} using conductivity detection. Traces of Co^{2+}, Zn^{2+} and Cd^{2+} were concentrated and separated with 10 mM tartaric acid at pH 2.54.

In another example a silica gel-based sorbent with chemically bonded amidoxime groups was used for chromatographic separation of transition and heavy metals [18]. The resin known as Amidoxim was obtained from Elsik, Russian Federation and had the structure:

$$Silica-O-Si-(CH_2)_2-\overset{\overset{\displaystyle NOH}{\|}}{\underset{\underset{\displaystyle NH_2}{|}}{C}}$$

Separation of five transition metals is shown in Fig. 7.15 using 5 mM sodium at pH 3.6 as the eluent. Post-column detection was employed with 0.5 mM 4-(2-pyridylazo) resorcinol (PAR) in 3 M ammonia and 1 M acetic acid as the color-forming reagent.

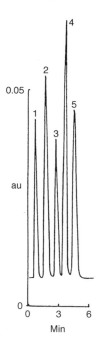

Figure 7.15. Separation of transition metals on the Amidoxim column with sodium oxalate as eluent. Eluent, 5 mM sodium oxalate (pH 3.6); other conditions as in Fig. 7.1. Peaks: 1 = Cd^{2+}; 2 = Co^{2+}; 3 = Zn^{2+}; 4 = Cu^{2+}; 5 = Ni^{2+}. From Ref. [18] with permission.

References

[1]   H. Small, T. S. Stevens and W. C. Bauman, Novel ion exchange chromatographic method using conductimetric detection, *Anal. Chem.*, 47, 1801, 1975.

[2]   J. S. Fritz, D. T. Gjerde and R. M. Becker, Cation chromatography with a conductivity detector, *Anal. Chem.*, 52, 1519, 1980.

[3]   S. Rabin, J. Stillian, V. Barreto, K. Friedman and M. Toofan, New membrane-based electrolyte suppressor device for suppressed conductivity detection in ion chromatography, *J. Chromatogr.*, 640, 97, 1993.

[4]   M. A. Rey and C. A. Pohl, Novel cation-exchange stationary phase for the separation of amines and of six common inorganic cations, *J. Chromatogr.*, A, 739, 87, 1996.

[5]   Dionex Corp., Ion pac CS12A cation exchange column. Fig. 6, 1995.

[6]   Dionex Corp., IonPac CS12A cation exchange column. Fig. 8, 1995.

[7]   V. Barreto, L. Bao, A. Bordunov, C., Pohl and J. Riviello, Electrolytic suppression in ion chromatography. Paper No. 958, Pittcon '99, Orlando, Fl.

[8]   P. R. Haddad and R. C. Foley, Aromatic bases as eluent components for conductivity and indirect ultraviolet detection of inorganic cations in non-suppressed ion chromatography, *Anal. Chem.*, 61, 1435, 1989.

[9]   A. Rahman and N. E. Hoffman, Retention of organic cations in ion exchange chromatography, *J. Chromatogr. Sci.*, 28, 157, 1990.

[10] N. E. Hoffman and J. Liao, Ion exchange in reversed-phase chromatography of some simple organic cations, *J. Chromatogr. Sci.*, 28, 428, 1990.

[11] P. J. Dumont, J. S. Fritz and L. W. Schmidt, Cation-exchange chromatography in nonaqueous solvents, *J. Chromatogr. A.*, 708–109–1995.

[12] P. J. Dumont and J. S. Fritz, Ion chromatographic separation of alkali metals in organic solvents, *J. Chromatogr. A*, 706, 149, 1995.

[13] G. J. Sevenich and J. S. Fritz, Addition of complexing agents in ion chromatography for separation of polyvalent metal ions, *Anal. Chem.*, 55, 12, 1983.

[14] G. J. Sevenich and J. Fritz, Effect of complexing agents on the separation of polyvalent cations, *J. Chromatogr.*, 347, 147, 1985.

[15] J. S. Fritz and T. A. Palmer, Ion exchange separations using sulfosalicylic acid, *Talanta*, 9, 393, 1962.

[16] J. Morris and J. S. Fritz, Ion chromatography of metal cations on carboxylic acid resins, *J. Chromatogr.*, 602, 111, 1992.

[17] G. Bonn, S. Reiffenstuhl and P. Jandik, Ion chromatography of transition metals on an iminodiacetic acid bonded stationary phase, *J. Chromatogr.*, 499, 669, 1990.

[18] I. N. Boloschik, M. L. Litvina and B. A. Rudenko, Separation of transition and heavy metals on an amidoxime complexing sorbent, *J. Chromatogr. A*, 671, 51, 1994.

[19] R. J. Phillips and J. S. Fritz, Synthesis and analytical properties of an N-phenylhydroxamic acid resin. *Anal. Chim. Acta*, 121, 225, 1980.

8 Ion-Exclusion Chromatography

8.1 Principles

Ion-exclusion chromatography (IEC) has developed into a very useful technique for separating relatively small weak acids (carbonic acid, carboxylic acids, hydrocarboxylic acids, etc.), weak bases (ammonia, amines) and hydrophilic molecular species such as carbohydrates and the lower alcohols. The analytical method actually involves the separation of molecular species rather than ions. Of course, ions can often be readily converted into molecular species as when anions of weak acids are acidified. The rationale for including IEC in a book on ion chromatography seems to be that a cation-exchange resin, or occasionally an anion-exchange resin, has generally been used for IEC separations. Also, it has become customary to include IEC in symposia and books devoted to ion chromatography.

Ion-exclusion chromatography is a comparatively old technique, attributed primarily to Wheaton and Bauman [1]. Ionic material is rejected by cation- or anion-exchange resin and passes through quickly, but non-ionic substances are held up and come through more slowly. Substances that can be separated include weak organic and inorganic acids, weak organic and inorganic bases, and hydrophilic neutral compounds such as sugars.

The resin bed consists of three parts:

(a) A solid resin network
(b) Occluded liquid within the resin beads
(c) The mobile liquid between the resin beads

The ion-exchange resin acts as a semipermeable membrane between the two aqueous phases, b and c. Ionized sample solutes are excluded from the interior water (b) and pass quickly through the column. Nonionic materials are not excluded and they partition between the two water phases, b and c. Thus, they pass more slowly through the column. Nonionic solutes differ in their degree of retardation by the resin phase because of: (1) differing polar attraction between the solute and resin functional groups, (2) differing van-der-Waal forces between the solutes and the hydrocarbon portion of the resin.

Harlow and Morman [2] studied the behavior of a large number of organic and inorganic acids on a column containing the hydrogen form of Dowex 50 × 12. The eluent was distilled water. The strongest acids, such as sulfuric and hydrochloric, elute together with the column void volume because they are highly ionized and cannot enter the resin phase. Harlow and Morman reported several generalizations for predicting elution behavior:

(1) Members of a homologous series emerge in order of increasing acid strength and decreasing water solubility. An example is the formic, acetic, and propionic acid series.

(2) Dibasic acids elute sooner than nonbasic acids. Oxalic acid elutes before propionic acid.

(3) An isoacid elutes before the corresponding normal acid. For example, isobutyric emerges before butyric acid.

(4) A double bond tends to retard the elution of an acid. Acrylic acid elutes after propionic acid.

(5) Acids with a benzene ring show a strong retention.

Tanaka and Ishizuka [3] studied the separation of acids by ion-exclusion chromatography on a modern, high performance cation-exchange column. Hitachi cation exchange resin, 8 % cross-linking, 18 ± 2 µm particle size was used. Retention volumes of acids were obtained at 50 °C using water as the eluent. These are given in Table 8.1.

Table 8.1. Retention volumes of acids. (Data from Ref. [3]).

Acid	Compound used	V_R (mL)	Acid	Compound used	V_R (mL)
HI	KI	12.8	HF	NaF	18.4
HBr	KBr	12.8	HCOOH	HCOOH	19.5
$HClO_4$	$HClO_4$	12.8	CH_3COOH	CH_3COOH, CH_3COONa	23.0
HCl	HCl, KCl	12.8	C_2H_5COOH		25.5
H_2SO_4	H_2SO_4, Na_2SO_4	12.8	C_3H_7COOH		30.0
HNO_3	HNO_3, KNO_3	12.8	H_2CO_3	Na_2CO_3	28.5
$C(OOH)_2$	$(COONa)_2$	13.0	HCN	KCN	28.5
H_3PO_3	$NaHPO_3$	13.8	H_3BO_3		28.5
H_3PO_2	NaH_2PO_2	14.0	C_6H_5OH	C_6H_5OH	28.2
H_3PO_4	H_3PO_4, Na_2HPO_4	14.2	CH_3OH	CH_3OH	28.8
H_2SO_3	Na_2SO_3	14.5	H_2S	Na_2S	34.8

These data suggest that the mechanism of elution is the same as for gel permeation chromatography, which follows the relationship:

$$V_R = V_o + K_d V_i \qquad (8.1)$$

In this equation, V_R is the retention volume, V_o is the column void volume (12.8 mL in this case), K_d is the distribution coefficient, and V_i is the "inner volume", which is the volume of water within the resin beads.

According to permeation theory, K_d should range from zero for compounds that are totally excluded from the resin to 1.0 for compounds that can completely permeate the resin. The results in Table 8.1 show that strong, highly ionized acids are all eluted in the void volume (12.8 mL). Several weak acids have almost identical retention volumes (average: 28.5 mL). If the value of K_d for these is assumed to be 1.0, V_i is then calculated to be 15.7 mL. Experimental measurement (by gravity) of V_i gave a value of 15.5 mL, which is in good agreement with the calculated value.

The retention volumes for butyric acid and H_2S are both higher than 28.5 mL, indicating that some additional retention mechanism is operative.

The acids in Table 8.1 with retention times between 12.8 mL and 18.5 mL can be assumed to have K_d values between zero and one. A linear plot was obtained for the pK, value of these acids vs retention time, thus demonstrating that retention is a function of the acid dissociation constant [3].

In addition to the permeation mechanism just described, sample compounds can be retained through interaction with the resin matrix. This type of interaction becomes stronger in larger organic molecules or in molecules with a group such as benzene ring which is attracted to the resin matrix. With such compounds, the retention times can usually be reduced by incorporating an organic solvent into the eluent.

8.1.1 Apparatus, Materials

IE chromatographic systems consist of the same components as any high-performance liquid chromatograph. In many cases, the resins, eluents, and detectors are similar to those used for ion chromatography. Stainless steel HPLC hardware is used except in cases where HCl eluents are used and plastic or glass parts must be used.

The most commonly used resins are gel type sulfonated cation exchangers or anion exchangers with a quaternary ammonium functional group. The exchange capacity of the resins used is generally higher than those used for ion chromatography. The resins may be either styrene–divinylbenzene copolymeric beads or polyacrylate beads. The diameter of the beads should be small and uniform. Resins with a 5 µm bead diameter are now available.

The extent of cross-linking of an ion-exchange resin plays a role in ion-exclusion chromatography. Generally the Donnan invasion of electrolytes is greater with resins of lower cross-linking. Greater penetration of a sample compound into the resin means that the retention time of that compound will be greater.

A wide variety of columns for IEC is available from Alltech, Dionex, Hamilton, Cetac, Tosoh and others. The column dimensions are often significantly larger (both wider and longer) than IC columns in order to separate substances with only small differences in K_d.

8.1.2 Eluents

Acidic eluents are used in separating weak acids (such as carboxylic acids) to repress their ionization and give sharp chromatographic peaks. However, water alone is often a suitable eluent for very weak acids such as carbonic acid and boric acid.

Organic amines require a basic eluent, such as dilute aqueous sodium hydroxide, to ensure that the amines are in the molecular form and are not ionized. Water alone can be used to elute weak molecular bases.

Sometimes an organic solvent is added to the aqueous eluent to speed up the elution of sample compounds. The type and concentration of the eluent must always be chosen so that it will be compatible with the detector. Methanol can shrink the bed of a sulfonated polystyrene catex resin and cause irreversible damage to the column. However, acetonitrile can be used up to 40 % (by volume) with this type of column. Ethanol and 2-propanol are acceptable in smaller amounts (about 15 % and 10 %, respectively) when they are used with the high-capacity cation exchanger.

8.1.3 Detectors

Conductivity, direct absorbance or a differential refractometer is the most common form of detection for IEC. Both non-suppressed and suppressed conductivity have been used extensively. The need to incorporate a low concentration of a strong acid into the eluent has been an impediment to direct conductivity detection.

Tanaka and Fritz [4] found that the conductivity detection of aliphatic carboxylic acids is much improved if a dilute solution of benzoic acid is used as the eluent. The background conductance is much lower than with aqueous solutions of strong mineral acids. A fairly low background conductance is obtained because benzoic acid is only partly ionized and a very dilute (5.0 mM) solution of benzoic acid is used. Nevertheless, this eluent is sufficiently acidic to give a good separation of aliphatic carboxylic acids with well-shaped peaks.

Membrane suppressors that can be regenerated continuously have also been developed for the ion-exclusion chromatography of weak acids. In IE, the primary contributor to the high eluent conductance is the hydrogen cation from the acid eluent. Membrane suppressors can reduce this background by exchanging the hydronium ion for tetrabutylammonium ion (TBA). This is shown in Table 8.2.

Table 8.2. Relative conductance of 1 mM solutions in suppressed ion-exclusion chromatography.

Ion Pair	Relative conductance, μS
H^+/Cl^-	425
TBA^+/Cl^-	100
TBA^+/OSA^-	40
$TBA^+/TDFHA^-$	30

TBA = tributylamine, OSA = octylsulfonic acid, TDFHA = tridecafluoroheptanoic acid.

Because of the TBA+/Cl- ion pair, the background conductivity is decreased significantly. It should be noted that the conductance of the sample species is reduced as well.

The background conductivity may be reduced further by replacing the HCl acid eluent with a weaker acid. Two acids that have been used successfully for IE chromatography are octylsulfonic acid (OSA) and tridecafluoroheptanoic acid (TDFA) [5].

8.2 Separation of Organic Acids

Gjerde and Mehra [6] compiled tables of retention factors for organic acids obtained on several commercial columns that are widely used for IEC.

Some interesting questions concerning the mechanism of IEC have been posed [7]. If the mechanism for separation of these carboxylic acids is primarily ion exclusion, why does an aliphatic carboxylic acid with a larger alkyl (R) group elute later than one with a smaller alkyl group? We would expect the K_d value of the more bulky acid to be lower than that of the less bulky acid. Again, why should substitution of an aliphatic hydrogen by a hydroxyl group cause the hydroxyl compound to be eluted faster (for example, tartaric acid is eluted before malic acid, and both are eluted before succinic acid)?

The reasonable answer seems to lie in the solute acid's interaction with the resin matrix. An acid with a large alkyl group would have a greater hydrophobic attraction and thus a longer retention time. A hydroxy compound is more polar than its hydrogen analogue and thus would interact less strongly with the resin matrix.

In IEC, stronger acids (lower pK_a) are eluted more rapidly than weaker acids, presumably because the stronger acids are incompletely converted to molecular form. If this were the case, coexistence of the ionic and molecular forms might produce broader peaks; however, these compounds produce very sharp peaks. A better explanation might be that stronger acids are more polar and therefore interact less strongly with the resin matrix.

These considerations suggest that differences in partitioning between the mobile and resin phases is the main mechanism for chromatographic separation of carboxylic acids and that an ion-exclusion mechanism is not essential. The separation of carboxylic acids with a water–acetonitrile gradient (Fig. 8.1) tends to support this conclusion. Other good separations were obtained [7] with an appropriately functionalized macroporous polymeric resin column. Because these resins are porous, they likely contain some stagnant mobile phase, but the amount of stagnant mobile phase is much less than that of gel resin columns and not nearly enough to account for retention of any of the carboxylic acid solutes. The separation mechanism seems to be a partitioning of solutes between the mobile phase and the resin. A nonfunctionalized polymeric resin can be used, but greater retention and better separations are obtained using a resin with polar substituents such as sulfonate or carboxylate. Good separations were obtained on a carboxylic acid resin column under conditions in which the resin's carboxylic acid group was in molecular rather than ionized form.

Separations are significantly faster on a macroporous resin column than on a gel resin column. Columns as short as 5 cm can be used with good results. Solvent gradients can be used with macroporous resins to obtain faster separations and sharper peaks for later eluted compounds. A data-acquisition system was used to correct a rising UV-absorbance baseline.

It has long been held that a separation of carboxylic acids by IEC requires the use of an acidic eluent to repress ionization of the analytes and thereby give sharp peaks. However, equilibrium constant calculations indicate that alkane carboxylic acids are

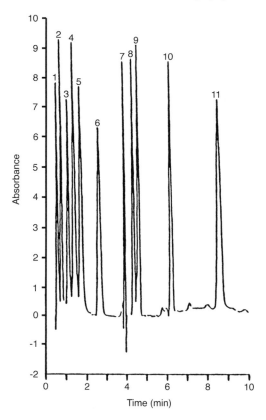

Figure 8.1. Gradient separation of carboxylic acids with background correction with a 15 cm × 4.1 mm Hamilton PRP-X300 column of 0.17 mequiv/g exchange capacity. Gradient: 1.0 mM sulfuric acid (pH 2.7) for 1.0 min, then a 0–20 % acetonitrile linear gradient over 4.0 min, followed by a 5.0 min hold at 20 % acetonitrile. Other conditions were the same as in Fig. 1. Peaks: 1 = oxalic acid, 2 = tartaric acid, 3 = maleic acid, 4 = citric acid, 5 = lactic acid, 6 = acetic acid, 7 = succinic acid, 8 = glutaric acid, 9 = propionic acid, 10 = butyric acid, 11 = valeric acid. From Ref. [7] with permission.

extensively ionized (60–97 %) in predominately aqueous solution at the low concentrations generally used in liquid chromatography. It is questionable that added sulfuric acid is really effective in converting the solute acids to their molecular form.

Morris and Fritz [8] studied the separation of low molecular weight carboxylic acids on a column (150 × 4.6 mm) packed with sulfonated PS-DVB resin spheres (0.25 mequiv/g exchange capacity). Elution with pure water gave fronted, poorly shaped peaks, as expected. However, incorporation of an organic solvent, but no acid, into the eluent gave an excellent separation of formic through valeric acid with sharp, well-shaped peaks. An optimal separation required 60 % methanol (40 % water v/v), 40 % ethanol, 20 % 1-propanol or 5 % 1-butanol. Conductivity detection could be used with very good sensitivity because none of the eluents contained any added acid. The sensitivity improved markedly as the organic alcohol content of the eluent decreased. Similar peak heights were obtained at the same detector setting for analytes at the following concentrations:

60 % methanol: 1 ppm formic acid, 25 ppm acetic acid, 30 ppm propionic acid, 50 ppm butyric acid, 75 ppm valeric acid.

40 % ethanol: 0.25 ppm formic to 6 ppm valeric acid.

20 % propanol: 0.25 ppm formic to 2 ppm valeric acid.

5 % butanol: 0.25 ppm formic to 0.50 valeric acid.

A separation in 5 % butanol is shown in Fig. 8.2.

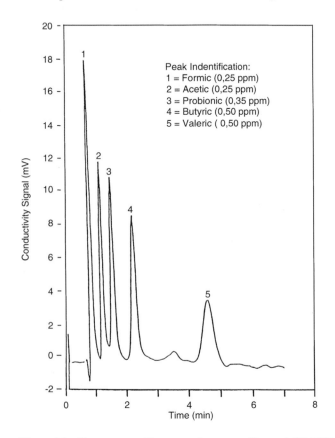

Peak Indentification:
1 = Formic (0,25 ppm)
2 = Acetic (0,25 ppm)
3 = Probionic (0,35 ppm)
4 = Butyric (0,50 ppm)
5 = Valeric (0,50 ppm)

Figure 8.2. Chromatographic separation on sulfonated PS-DVB resin column (150 × 4.6 mm) of 0.25 mequiv/g exchange capacity. Eluent conditions: 5 % butanol/deionized water. Detection is conductivity with an output range of 3 μS full scale. Flow rate is 1.0 mL/min. Peaks are as identified. From Ref. [8] with permission.

8.2.1 Mechanisms of Alcohol Modifiers

The effect of methanol and ethanol on the separations could possibly be explained by stronger solvation of the sample solutes in the mobile phase. However, this explanation becomes unlikely for an eluent containing only 5 % 1-butanol, the remainder being water.

Scott and Simpson studied the adsorption of aliphatic alcohols, aldehydes, and carboxylic acids in binary mixtures with water by ODS-2 silica [9]. They found that the distribution coefficient increases exponentially with the carbon number of the moderator. When using an aliphatic moderator having a chain length of four or five carbon atoms, the surface of a bonded phase could be completely covered with a monolayer. They stated further that the chromatographic characteristics of the surface could be changed by choosing appropriately active groups.

Adsorption of a layer of alcohol on the polymeric resin surface is believed to explain the dramatic effects observed in our separations of carboxylic acids. Butanol has the highest distribution coefficient of the alcohol moderators studied, and only a low concentration in the aqueous eluent is needed to coat the resin surface. Partitioning of the various solute acids between the predominately aqueous eluent and the coated resin surface is much different that it is with an uncoated polystyrene surface.

In an effort to gain further insight as to the retention mechanism, sodium salts of the carboxylic acids were injected rather than the acids themselves [8]. The chromatograms were almost identical, even with regard to peak height. At first it was assumed that the sulfonic acid groups on the resin converted with sodium salt to the molecular acid. But running the separations on unsulfonated PS-DVB resins still gave comparable chromatograms for the carboxylic acids and their sodium salts. In reviewing the complete chromatographic system no source of acid other than carbonic acid (or CO_2) could be found. Since the analytes are moderately weak acids, it is possible that carbonic acid could have an effect on the retention mechanism. It is known from previous work that carbonic acid has a substantial retention factor and thus spends a significant amount of time residing on the surface of the polymeric resin.

It was demonstrated that carbon dioxide or carbonic in the 1-butanol-water eluent equilibrates with the resin to form an adsorbed layer that is responsible for conversion of carboxylate salts to the molecular form.

Adsorbed $H_2CO_3 + RCO_2^- \rightarrow$ Adsorbed $HCO_3^- + RCO_2H$

The adsorbed carbonic acid also affects the separations. Figure 8.3A shows the separation of four carboxylic acids after equilibration of the column with 5 % 1-butanol eluent. Then the column was re-equilibrated with eluent that was freed of carbonic acid by placing an anion-exchange column in the hydroxide form in-line between the pump and injection valve. After 30 min, the background conductance dropped from approximately 900 nS to 180 nS and the separation in Fig. 8.3B was incomplete. After switching back to the original eluent the adsorbed carbonic acid layer gradually reformed, giving the chromatogram in Fig. 8.3C. It was concluded that the presence of carbonic acid in the eluent appears beneficial for the separation of carboxylic acids so long as it is in moderate and controlled quantities.

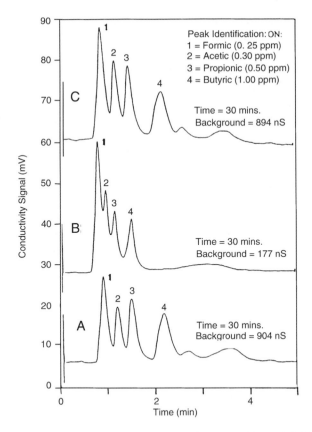

Figure 8.3. Chromatographic separation on sulfonated PS-DVB resin column (150 × 4.6 min) of 0.25 mequiv/g exchange capacity. (A) initial separation, (B) separation after 30 min with anion-exchange column (hydroxide form) in-line between pump and injection valve, and (C) separation 30 min after removal of anion-exchange column (initial configuration). Eluent conditions: 4 % butanol/deionized water. Detection is conductivity with an output range of 3 μS full scale. Flow rate is 1.0 mL/min. Peaks are as identified. From Ref. [8] with permission.

8.3 Determination of Carbon Dioxide and Bicarbonate

The determination of carbon dioxide is an important analytical problem, especially when low concentrations are to be measured. Ion-exclusion chromatography provides a convenient way to determine carbon dioxide, or its form in solution which is molecular carbonic acid. The separation column is packed with a cation exchange resin in the H^+ form so that salts are converted to the corresponding acid. Ionized acids pass rapidly through the column while molecular acids are held up to varying degrees. A conductivity detector is commonly used.

Unfortunately, carbonic acid is a very weak acid ($pK_1 = 6.4$) and the conductance of the carbonic acid peak consequently is very low. In order to obtain a more sensitive detection, Tanaka and Fritz inserted a cation-exchange column in the K^+ form between the cation-exchange column and the detector [10]. Its purpose is to convert the carbonic acid to a more highly ionized form and thereby to increase the conductivity. This is called the first enhancement column. When it is in the K^+ form, the exchange reaction is as follows:

$$R_s-K^+ + H_2CO_3 \rightarrow R_s-H^+ + K^+ + HCO_3^- \qquad (8.2)$$

With this column in place a standard sample of 1.0 mM bicarbonate gave a conductance of 0.504 µS, which was approximately 5.5 times greater than that obtained with no enhancement column.

8.3.1 Enhancement Column Reactions

A calculation of the relative detector signals shows that the conversion of carbonic acid to potassium bicarbonate and then to potassium hydroxide in the enhancement columns is essentially quantitative. In particular, it may seem surprising that an acid as weak as carbonic acid ($k_{1a} = 4.0 \times 10^{-7}$) is able to exchange its H^+ for K^+ on the resin.

$$H_2CO_3 + R_s^- K^+ \rightleftarrows R_s^- H^+ + K^+ + HCO_3^- \qquad (8.3)$$

However, two points should be kept in mind. One is that an increasing fraction of carbonic acid is ionized as the solution becomes more dilute (~6.8 % in 0.1 mM carbonic acid, for example). A second point is that the high concentration of K^+ on the exchange column (~4.2 M) pushes the ion-exchange equilibrium to the right. For 0.1 mM carbonic acid, it can be calculated that only a few theoretical plates would be needed for complete conversion of H_2CO_3 to K^+ and HCO_3^-.

Strong acids will also undergo ion exchange in the enhancement columns. Since the eluent is simply water, the enhancement columns will last a long time before regeneration or replacement is necessary unless the sample analyzed have a very large amount of strong acids or their salts. It is possible that ion-exchange membrane reactors, which can be continuously regenerated, could be used in place of the enhancement columns.

A second enhancement column, placed just after the first enhancement column, provides still more sensitive detection. This is an anion-exchange column in the hydroxide form that converts $K^+HCO_3^-$ (equivalent conductance 118) to K^+OH^- (equivalent conductance 272).

The chromatograms in Fig. 8.4 show the effect of enhancement columns. Each enhancement column increases the carbonic acid peak height, but the baseline conductance is also increased somewhat. A pre-column packed with anion-exchange resin in the OH^- form was then placed between the pump and loop injector to remove completely and continuously the carbon dioxide in the eluent. This arrangement resulted in a significant decrease in eluent background conductance, as shown in Fig. 8.4D. An almost linear calibration plot was obtained from 0.05 to 5.0 mM bicarbonate. The detection limit was estimated to be 1.45 µM.

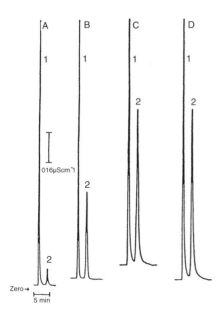

Figure 8.4. Comparison of ion-exclusion chromatograms of HCO_3^- with and without first and second enhancement columns and precolumn: (A) separating column alone (H^+ form) (no enhancement); (B) separating column (H^+ form) + first enhancement column (K^+ form); (C) separating column (H^+ form) + first enhancement column (K^+ form) + second enhancement column (OH^- form); (D) (C) with precolumn (OH^- form) for removal of CO_2 gas in water eluent. Conditions: first enhancement column is TSK SCX 5 μm, 4.6 × 50 mm. Separating column is TSK SCX (H^+ form) 5 μm, 7.5 × 100 mm. Eluent is water (1 mL/min). Sample mixture of 1 mM KCl and 1 mM $NaHCO_3$ (0.1 mL). From Ref. [10] with permission.

8.4 Separation of Bases

Haddad et al. measured retention volumes for a variety of bases on a quaternary ammonium functionalized PS–DVB stationary phase using dilute aqueous sodium hydroxide as the eluent [11]. Values for the retention volumes and distribution coefficients of selected bases are given in Table 8.3. Strong bases, which are fully ionized a the eluent pH, elute at the column void volume and have a K_d value of 1.0. Solutes intermediate between these two extremes are partly ionized and generally can be separated by an ion-exclusion mechanism.

Higher aliphatic amines (butylamine, pentylamine, diethylamine, etc.) had larger retention volumes and K_d values well above 1.0. A mixed retention mechanism involving hydrophobic adsorption and steric effects was observed for these compounds. Aromatic amines were found to be retained almost solely by a reversed-phase mechanism involving interaction of the solute with the unfunctionalized regions of the stationary phase. Retention of these solutes could be manipulated most easily by addition of acetonitrile to the eluent.

Table 8.3. Retention data for basic compounds. Bio-Rad (Richmond, CA, USA). 300 × 7.8 mm column containing quaternary-ammonium PS-DVB resin. Column void volume 3.8 mL; sum of dead and the inner column volumes, 10.3 mL. (Data from Ref. [11]).

Solute	pK_2	V_R (mL)	K_d
KOH	−10.00	3.90	0.02
NaOH	−5.00	3.90	0.02
Ca(OH)$_2$	2.43	4.00	0.03
Ethylenediamine	4.07	5.30	0.23
Hydrazine	5.77	6.20	0.37
Methylamine	3.34	6.60	0.43
Ammonia	4.75	6.96	0.49
Triethanolamine	6.24	7.00	0.49
Ethylamine	3.30	7.15	0.52
Trimethylamine	4.19	7.36	0.55
Propylamine	3.40	8.00	0.65
Diethylamine	2.96	8.60	0.74
Methanol	15.00	10.30	1.00
Urea	13.82	10.32	1.01
Thiourea	14.26	10.40	1.02

8.5 Determination of Water

If an alcohol such as methanol can be separated by ion-exclusion chromatography using water as the eluent, why not do the reverse and separate water using a methanol eluent? Stevens et al. [12] did just this. They added a small amount of sulfuric acid to the eluent and detected the chromatographic water peak by a decrease in conductivity. The main drawback with this method was a non-linear calibration curve with very poor detection sensitivity in some concentration regions.

Fortier and Fritz [13] separated water by IEC and devised a unique equilibrium system for in-line spectrophotometric detection. This method has been refined and its capabilities expanded by continuing research by Chen and Fritz [14–17]. Water is separated chromatographically from the other sample components on a short column packed with cation-exchange resin in the H$^+$ form using dry methanol as the eluent. Detection of the water peak is made possible by addition of a low concentration of cinnamaldehyde to the methanol eluent. In the presence of an acid catalyst, such as a H$^+$-cation exchanger, cinnamaldehyde reacts with methanol to form the dimethylacetal.

$$(+ H^+)$$
$$C_6H_5CH = CHCHO + 2\ CH_3OH\ H^+ \rightleftarrows C_6H_5CH = CHCH(OCH_3)_2 + H_2O \quad (8.4)$$

The UV spectra of cinnamaldehyde and its reaction product (an acetal) are quite different, as shown by Fig. 8.5. The necessity of an acid catalyst must be emphasized. Cinnamaldehyde dissolved in methanol will retain its own spectrum (A in Fig. 8.5) for

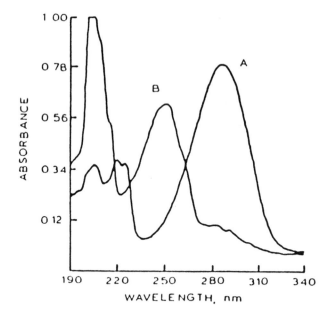

Figure 8.5. Spectra of 0.0318 mM *trans*-cinnamaldehyde in methanol. (A) Spectrum immediately after the solution was prepared; (B) spectrum after the solution had been shaken with Aminex Q-150S in the H$^+$ form. From Ref. [13] with permission.

some time but introduction of an acid catalyst results in a very fast equilibration to form the acetal. This is accomplished by the use of a sulfonic acid cation exchanger for the chromatographic separation. Since most of the cinnamaldehyde has been converted to the acetal, the background absorbance at 300 nm is low. However, a water zone passing through the column will shift the equilibrium towards the formation of more cinnamaldehyde and the absorbance at 300 nm will increase.

$$H_2O + acetal \overset{H^+}{\rightleftarrows} aldehyde + 2\ CH_3OH \tag{8.5}$$

In methanol the equilibrium constant, K, has been measured:

$$K = [aldehyde]/[acetal][H_2O] = 5.3 \times 10^{-4} \tag{8.6}$$

The detector signal (A_{det}), which is the change in absorbance when water passes through the detector cell is given by the following equation:

$$A_{det} = k\ C_{ca}(C_{samp} - C_{blank}) \tag{8.7}$$

where k is a proportionality constant related to the equilibrium constant K, C_{ca} is the total concentration of cinnamaldehyde added to the eluent, C_{samp} is the water concentration of sample, and C_{blank} is the water concentration in the eluent itself. As predicted by this equation, the detector signal has been shown experimentally to be a linear function of the total aldehyde and of the concentration of water present.

Typically, a sharp water peak is obtained in approximately 2 min. The water peak is always well separated from an earlier injection peak that is due to the sample matrix. Under favorable conditions a very short column (length 2.5 cm) can be used and a water peak obtained in as little as 20 s [17].

Determination of water in sample containing aldehydes or ketones has always been a problem. These compounds can react with a solvent (methanol) to produce an acetal or ketal plus water.

$$RCHO + 2 \ CH_3OH \rightleftarrows RCH(OCH_3)_2 + H_2O \tag{8.8}$$

The key to this problem is that the above reaction will not take place unless H^+ is present to catalyze the reaction. By using a cation-exchange column in the Li^+ form, water in the acetone can be separated chromatographically from the acetone. A H^+-form column placed in series then catalyzes the cinnamaldehyde–acetal equilibrium shift that is necessary for detection of the water. Reaction with methanol to form water is also catalyzed in this second column, but separation of the acetone and initial water has already taken place in the first column.

The system described here is an example of a post-column reaction system which uses a solid-phase reactor. The set-up is simple and works very well. The reactants are already present in the mobile phase. The reaction simply does not occur until the catalyst column is reached. No additional reagents are mixed with the effluent stream.

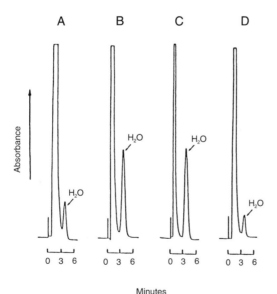

Figure 8.6. Determination of water in various aldehydes (A) 0.11 % water in acetaldehyde, (B) 1.57 % water in propionaldehyde, (C) 0.81 % water in heptaldehyde, and (D) 0.19 % water in octylaldehyde. From Ref. [14] with permission.

There is no need for the additional hardware (second pump, mixing tee or reaction chamber) commonly used in post-column reaction systems. Consequently, the problems inherent in a typical post-column reaction system are avoided.

Chromatograms for determination of small amounts of water in several aldehydes are shown in Fig. 8.6. This same "two-column" method is also useful for other difficult samples. For example, peroxides are highly oxidizing and interfere with the Karl Fischer titrimetric determination of water. Water has been determined by the chromatographic method in several organic peroxides [17].

8.5.1 Determination of Very Low Concentrations of Water by HPLC

The inability to obtain really dry methanol limits the ability of the liquid chromatographic (LC) method to determine very low concentrations of water. Attempts to remove water from methanol by treatment with molecular sieves or distillation from calcium hydride still gave a product with at least 50 to 150 ppm water.

Virtually all the water can be removed from methanol by adding an ortho ester, trimethylorthoformate (TMOF), and a small amount of sulfuric acid to catalyze the reaction [17].

$$CH(OCH_3)_3 + H_2O \xrightarrow{H^+} HCO_2CH_3 + 2\ CH_3OH \tag{8.9}$$

The amount of TMOF needed to react with the water in the methanol eluent (containing cinnamaldehyde and a low concentration of sulfuric acid) is determined by a titration procedure. At first, each addition of TMOF solution to the eluent produces a decrease in the detector signal as the eluent is pumped through the chromatographic UV–VIS detector. As the water concentration in the methanol becomes progressively lower, the detector signal changes less and less. TMOF is added just until there is no further lowering of the detector signal.

Removal of almost all the water from the methanol reduces the eluent baseline considerably and also increases the height of the water peak for a sample. With this treatment the detection limit is estimated to be <5 ppm water.

8.6 Simultaneous Separation of Cations and Anions

Acid rain caused by SO_2 and NO_x in air is a major environmental pollution problem in many parts of the world. The major cationic components of acid rain are H^+, Na^+, NH_4^-, K^+, Mg^{2+} and Ca^{2+}; the major anionic components are Cl^-, NO_3^- and SO_4^{2-}. The ionic balance between the total positive charge and negative charge of these ions is almost 100 %, so the simultaneous determination of these ions is important.

Tanaka et al. devised a method for determination of the ions in acid rain in a single run [18]. The column used contained a polyacrylate weak-acid cation exchange resin (TOSOH TSK gel OA-PAK, 300 × 7.8 mm, 5 µm particle size). Water alone as the mobile phase only separated weak acid anions from strong acid anions as a group; cations remained fixed on the column. When the aqueous mobile phase contained sulfuric acid, it was possible to separate the cations by cation exchange but no separation of the anions was obtained. An eluent containing tartaric or citric acid ($pK_1 \approx 2–3$) made it possible to separate simultaneously both the cations and anions in acid rain. The optimized eluent contained 5 mM tartaric acid and 7.5 % methanol.

A separation of the ions in acid rain is shown in Fig. 8.7 using conductivity detection. The peaks of the anions (sulfate, chloride and nitrate), which are highly ionized, are positive. The cation peaks are of lower conductivity than the tartaric acid eluent and hence are in the negative direction. The detection limits are low enough to handle most acid rain samples without any preconcentration (Table 8.4).

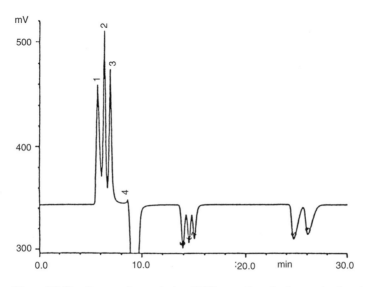

Figure 8.7. Simultaneous ion-exclusion-CEC separation of anions and cations by elution with 5 mM tartaric acid/7.5 % methanol-water at 1.2 mL/min. Eluent conductivity: 536 µS cm^{-1}. Peaks: 1 = SO$_4^{2-}$; 2 = Cl$^-$; 3 = NO$_3^-$; 4 = eluent dip; 5 = Na$^+$; 6 = NH$_4^+$; 7 = K$^+$; 8 = Mg^{2+}; 9 = Ca^{2+}. From Ref. [17] with permission.

Table 8.4. Detection limits of major anions and cations related to acid rain water determined by elution with 3 mM tartaric acid/7.5 % methanol–water.

Ion	Detection limit[a]	
	µM	ppb
SO$_4^{2-}$	0.16	15
Cl$^-$	0.10	3.6
NO$_3^-$	0.14	9
Na$^+$	0.20	4.6
NH$_4^+$	0.30	5.4
K$^+$	0.32	12.5
Mg^{2+}	0.28	6.8
Ca^{2+}	0.38	11.2

[a] Signal-to-noise ratio = 3

The mechanism for separation of sulfate, chloride and nitrate is not entirely clear. Anions that are completely ionized normally cannot be separated by an ion-exclusion process. A weak hydrophobic effect might account for the slight differences in retention of these anions.

8.7 Separation of Saccharides and Alcohols

Although not strictly ion chromatography, the separation of carbohydrates are discussed here for two reasons. First, the columns and eluents used for the separations are the same or are quite similar to those used for weak organic and inorganic acids. Second, the samples often contain both carbohydrates (sugars) and organic acids, and it is desirable to perform the analysis in the same run. Wines, for example, contain organic acids, residual sugars, and alcohols and can be analyzed using ion-exclusion type columns. Mono-, di-, and polysaccharides, sugar alcohols and organic acids are often found together in food and drink.

8.7.1 Separation Mechanism and Control of Selectivity

The basic mechanism of separation of carbohydrates is by ligand exchange chromatography but is quite similar to ion-exclusion chromatography described earlier in this chapter for weak organic and inorganic acids. The column contains fully sulfonated polystyrene polymer beads cross-linked with polydivinylbenzene. The polymers are fully hydrated and contain occluded water within the gel polymer matrix, just as in ion-exclusion polymer beads. Analytes partition between the occluded water within the bead matrix and the mobile phase. Water is most often used as the mobile phase and the detection method is most often refractive index.

Unlike many chromatographic methods, carbohydrate, ligand exchange columns are most often operated with eluents that do not contain mobile phase modifiers, organic solvents or counterions. Since the eluent is water, the column packing itself must be modified in order to change or control the separation selectivity. The most common means of making these changes is by changing the ionic form of the polymer packing.

In ligand exchange chromatography, metal ions are ionically bound to the polymer through the sulfonate ion exchange group. The column selectivity for the various carbohydrates depends on the relative attraction of water molecules to the metal ions that are ionically bound to the polymer and the hydroxyl groups of the saccharides. Changing the ionic form of the column from hydrogen to sodium and then to calcium form generally increases retention times and improves resolution for most carbohydrates. The hydroxyl groups on the carbohydrate form a more stable complex with the metal resulting in an increasing retention. Changing to lead or silver has an even more dramatic effect, but may also result in unnecessary long analysis times. Table 8.5 shows the effect of ionic form of the polymer.

The calcium-form columns are the most popular and are usually the column of choice for most separations. Figure 8.8 shows a typical separation. Some separations

Table 8.5. Comparison of metal ionic form and cross-linking. Data courtesy of Transgenomic.

Column Name	Retention Time (min) CHO611	CHO620	CHO682	COR87H	COR87N	COR87K	COR87C	COR87Pb
ionic form	Na	Ca	Pb	H	Na	K	Ca	Pb
% cross-linking	6	6	6	8	8	8	8	8
Compound:								
Arabinose	11.08	10.64	23.95	12.08	12.64	14.72	13.92	16.32
Digitoxose	10.18	10.26	21.95		11.40	12.32	14.19	15.48
Fructose	10.33	10.07	25.84	11.25	11.61	13.31	13.63	16.96
Fucose	10.96	10.57	24.16	12.80	12.34	14.39	13.82	16.44
Galactose	10.22	9.58	22.32	11.12	11.44	13.36	13.82	15.16
Glucose	9.53	8.72	19.14	10.57	10.72	12.55	11.17	13.38
Mannose	10.27	9.79	25.50	11.13	11.57	13.74	12.76	16.76
Rhamnose	9.88	9.64	22.56	11.94	11.08	12.83	12.86	15.26
Sorbose	9.33	9.50	22.38	10.08	11.08	12.66	12.86	15.24
Tagatose	10.29	11.53		11.15	11.36	12.82	16.46	20.80
Xylose	10.34	9.56	20.64	11.32	11.77	13.69	12.32	14.42
Cellobiose	7.17	6.65	15.58	8.43	7.90	9.26	8.94	10.98
Lactose	7.51	7.01	17.37	8.77	8.18	9.63	9.44	11.84
Lactulose	7.85	7.57	20.70	9.00	8.48	10.08	10.17	13.24
Melibiose	7.46	6.99	17.63	8.56	8.19	9.72	9.36	12.02
Trehalose	7.14	6.70	15.98	8.64	7.85	9.02	9.07	11.20
Sucrose	7.27	6.76	15.70		7.99	9.11	9.09	11.10
Maltose	7.37	6.89	16.61	8.57	8.08	9.48	9.17	11.54
Ribitol	10.13	10.94	30.72	12.44	11.26	11.84	15.55	20.44
Arabitol	10.52	12.32	39.82	12.65	11.64	12.10	18.36	25.24
Galactitol	10.23	13.05	52.43	11.80	11.15	11.61	20.46	31.60
Myo-Inositol	11.01	10.82	35.58	11.02	12.48	14.08	14.27	20.06
Lacititol	7.87	8.55	33.23	9.26	8.45	9.34	12.17	19.50
Malititol	7.68	8.54	30.38	9.00	8.28	9.06	12.22	17.76
Mannitol	9.90	11.84	40.03	11.66	10.81	11.42	17.81	24.98
Sorbitol	10.38	13.64	56.56	11.77	11.32	11.86	21.34	33.40
Xylitol	11.01	13.93	51.15	12.82	12.16	12.64	21.30	31.10
Amiprylose	4.20	4.50		6.86	5.74	6.42	7.68	9.46
Melezitose	6.01	5.78	13.85		6.81	7.82	8.20	13.08
Maltotriose	6.22	5.91	15.17	7.72	6.98	8.16	8.28	10.54
Raffinose	6.10	5.86	14.40		6.88	7.92	8.24	10.22
Stachyose	5.39	5.28	13.41		6.33	7.28	7.77	9.58
Maltotetrose	5.54	5.37	14.07	7.30	6.42	7.46	7.80	9.84
Maltopentose	5.08	5.00	13.08	7.10	6.11	7.02	7.53	9.34
Maltohexose	4.87	4.78	12.24	7.00	5.94	6.74	7.38	8.80
Maltoheptose	4.60	4.66	11.74	6.96	5.84	6.61	7.28	8.52
Nitrate	4.20	4.50	10.30	6.85	5.70	6.40	7.30	8.40

Flow Rate: 0.5 mL/minute; Temperature: 90 °C

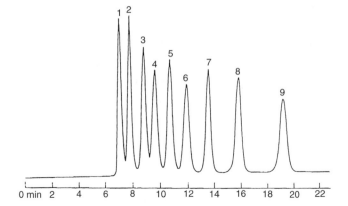

Figure 8.8. Separation of various sugars and sugar alcohols on a Coregel-87C column. Analysis Conditions: Column: Transgenomic Coregel-87C; Eluent: Distilled water; Flow rate: 0.6 mL/min; Temperature: 85 °C; Pressure: 425 psig; Detection: Rl range 8x; Injection: 20 μL. Sample: 1. raffinose, 2. sucrose, 3. lactulose, 4. glucose, 5. galactose, 6. fructose, 7. ribitol, 8. mannitol, 9. sorbitol.

require resin to be in different metal ionic forms. Arabinose and mannose, for example, are quite difficult to separate except with a lead-form column. Lead columns are used for mono-, di-, tri-, and tetra-saccharides and for sugar alcohols.

Hydrogen form columns are also useful when mixtures of sugars, acids and alcohols are to be separated. Ion exclusion or ligand exchange are both appropriate names for this type of chromatography because both separation mechanisms apply. Figure 8.9 shows a separation of a wine sample. Because the column is in the hydrogen form, it is possible to use an eluent containing an acid without fear of changing the ionic form. The pH of the eluent will control the selectivity of the organic acid analyte, but gener-

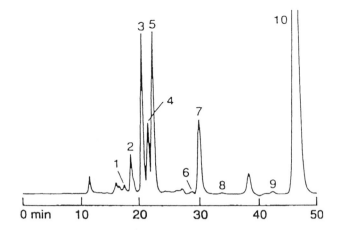

Figure 8.9. Wine analysis by high resolution ion exclusion. Analysis Conditions: Column: Transgenomic ION-300; Eluent: 0.005 N H_2SO_4; Flow rate: 0.3 mL/min; Temperature: 60 °C; Detection: Drl; Sample: 1. citric acid, 2. tartaric acid, 3. glucose, 4. malic acid, 5. fructose, 6. acetic acid, 7. glycerol, 8. methanol, 9. ethanol.

ally will not affect the sugars or the alcohols. Table 8.6 shows that lowering the eluent pH will retain organic acids more strongly, but will not affect sugars or alcohols. Generally, increasing column temperature will increase organic acid ionization and will decrease retention of those analytes.

Table 8.6. Retention times of acids, sugars and alcohols on a Transgenomic COR87H ion-exclusion column. Eluent: H_2SO_4, 0.01N, 0.003N, and 0.001N. Flow Rate: 0.6 mL/min. Temperature: 25°C. Detection:RI.

Substance	Retention Time (min)			Substance	Retention Time (min)		
	0.01 N	0.003 N	0.001 N		0.01 N	0.003 N	0.001 N
	pH 2.1	pH 2.5	pH 3.0		pH 2.1	pH 2.5	pH 3.0
Oxalic acid	6.4	6.2	6.1	Ascorbic acid	9.9	9.8	9.5
Sucrose	6.8	6.9	6.8	Malonic acid	9.9	9.8	9.4
cis-Aconitic acid	7.1	9.8	8.4	Glyceric acid	10.5	10.5	9.9
Glucutonic acid	7.3	7.2	6.9	*t*-Aconitic acid	11.2	10.3	9.0
Oxaloacetic acid	7.5	7.5	6.7	Glucotonic acid γ-Lactone	11.2	11.2	11.0
Citric acid	7.8	7.6	7.2	Glycerol	12.1	12.5	12.4
Isocitric acid	8.0	7.8	7.5	Lactic acid	12.1	12.1	11.8
α-Ketoglutaric acid	8.2	7.5	6.8	Succinic acid	12.1	12.1	11.8
Glucose	8.3	8.3	8.2	Shikimic acid	12.2	12.4	12.0
Ginconic acid	8.3	8.1	7.6	Formic acid	13.6	13.5	13.1
Galacturonic acid	8.3	8.1	7.6	Barbituric acid	13.9	13.5	13.5
Gluconic acid δ-Lactone	8.4	8.4	7.9	Acetic acid	14.6	14.7	14.6
Maleic acid	8.4	7.2	7.2	Glutaric acid	15.0	14.9	14.7
Tartaric acid	8.5	8.3	7.7	Fumaric acid	16.8	15.4	13.6
Isocitric acid γ-Lactone	8.7	7.4	7.2	Methanol	16.9	17.2	17.2
Pyruvic acid	8.9	7.9	7.4	Propionic acid	17.3	17.3	17.2
Glyoxillic acid	9.1	8.9	8.6	Adipic acid	18.8	18.5	18.2
Fructose	9.2	9.3	9.1	Ethanol	18.8	18.7	18.8
Citramalic acid	9.4	9.3	8.7	Acetone	21.3	21.2	21.4
Malic acid	9.5	9.5	8.9	*n*-Butyric acid	21.4	21.5	21.3
Quinic acid	9.8	9.7	9.2	*o*-Phthalic acid	34.0	29.1	23.1

There are other subtle changes that can take place in column manufacturing that will provide differences in column performance. Changes in cross-linking density, microporosity, ion exchange capacity, and particle size all will affect the column performance. The beads are sulfonated to a capacity of 3–5 mequiv/g (1–2 sulfonic acid groups are present per polymer aromatic group). The polymeric beads used for these separations are low cross-linked gel-like material. A higher cross-linked bead is more mechanically durable. However, the selectivity may not be suitable for separation. Generally lower cross-linked gels give the best separations, but they are the most fragile. The column bed can collapse through high pumping pressures. Once the bed starts to collapse, the feedback mechanism is positive leading to higher eluent backpressures and further collapsing of the column bed. By far the most fragile column is the Pb column because this column is usually very low cross-linked. One should not attempt to pump eluent through this or any carbohydrate column without first heating the column. It may also be necessary to elevate the temperature of the column to eliminate separation of the various stereoisomers of monosaccharides. Besides keeping the backpressure of the eluent low, the higher eluent temperatures usually improves the separation. The optimum temperature is usually between 60° and 85° C, and one should experiment to find the best conditions for each particular analysis. On the other hand, if chiral separation for sugars is desired, lowering column temperature to ambient and below slows mutarotation and allows resolution of many isomers.

8.7.2 Detection

Refractive index is the most common detector because the samples are not concentration limited (there is plenty of fructose in corn syrup, for example) and refractive index is a universal detector. UV detection is common when the analyte is UV absorbing. If high sensitivity is needed, pulsed amperometric detection (PAD) may be used to detect carbohydrates after separation by ion exchange. The base that is used to ionize the sugars also serves as a medium to allow oxidation of the sugars by PAD (see Chapter 4). Sugars will not oxidize so easily and be detected unless the pH is basic. Extra base is often added for PAD to work for ion exchange separations of carbohydrates (Dionex Technical Note 20). The same can be done for PAD detection of carbohydrates after separation by ion exclusion. The concentration of sodium hydroxide needed for PAD detection is dependent on the detector setting, but are in the range of 15–900 mM (Dionex Technical Note 21).

8.7.3 Contamination

Because of differences in swelling, columns are converted to the appropriate metal form in the bulk form before packing the column. Samples containing salts (cations) should be desalted to prevent displacement and precipitation of the metal from the column. If a metal gets displaced by other metals that may be in the sample or eluent, it is sometimes possible to regenerate the column with salt. For example, CaEDTA can sometimes be used to regenerate a Ca-form column. This regeneration should be done in the backflush mode.

It is sometimes possible to remove organic contamination from a column by pumping a 40 % acetonitrile solution in the backflush mode at a very low flow rate (0.1 mL/min). This should be done only as a last resort to recover a column. The use of other types of organic solvents is not advised because bed shrinkage may result.

References

[1] R. M. Wheaton and W. C. Bauman, Ion exclusion, *Ind. Eng. Chem.*, 45, 228, 1953.

[2] G. A. Harlow and D. H. Morman, Automatic ion exclusion-partition chromatography of acids, *Anal. Chem.*, 36, 2438, 1964.

[3] K. Tanaka and T. Ishisuka, Elution behavior of acids in ion-exclusion chromatography using a cation-exchange resin, *J. Chromatogr.*, 174, 153, 1979.

[4] K. Tanaka and J. S. Fritz, Separation of aliphatic carboxylic acids by ion exlusion chromatography using a weak-acid eluent, *J. Chromatogr.*, 361, 151, 1986.

[5] Dionex Corp., Application note 25, Sunnyvale, CA, 1980.

[6] D. T. Gjerde and H. Mehra, Advances in ion chromatography, P. Jandik and R. M. Cassidy Eds, Century International, Medfield, MA, Vol. 1, p139, 1989.

[7] J. Morris and J. S. Fritz, Separation of hydrophilic organic acids and small polar compounds on macroporous columns, *LC-GC*, 11, 513, 1993.

[8] J. Morris and J. S. Fritz, Eluent modifiers for the liquid chromatographic separation of carboxylic acids using conductivity detection, *Anal. Chem.*, 66, 2390, 1994.

[9] R. P. W. Scott and C. F. Simpson, Solute-solvent interactions on the surface of reversed phase, *Faraday Symp. Chem. Soc.*, 15, 69, 1980.

[10] K. Tanaka and J. S. Fritz, Determination of bicarbonate by ion-exclusion chromatography with ion-exchange enhancement of conductivity detection, *Anal. Chem.*, 59, 708, 1987.

[11] P. R. Haddad, F. Hao and B. K. Glod, Factors affecting retention of basic solutes in ion-exclusion chromatography using an anion-exchange column, *J. Chromatogr. A.*, 671, 3, 1994.

[12] T. S. Stevens, K. M. Chritz and H. Small, Determination of water by liquid chromatography using conductometric detection, *Anal. Chem.*, 59, 1716, 1987.

[13] N. Fortier and J. S. Fritz, Use of a post-column reaction and a spectrophotometric detection, *J. Chromatogr.*, 462, 325, 1989.

[14] J. Chen and J. S. Fritz, Chromatographic determination of water using spectrophotometric detection, *J. Chromatogr.*, 482, 279, 1989.

[15] J. S. Fritz and J. Chen, New chromatographic methods for the determination of water, *Am. Lab.*, 24J, July, 1991.

[16] J. S. Fritz, Principles and applications of ion-exclusion chromatography, *J. Chromatogr.*, 546, 111, 1991.

[17] K. Tanaka, K. Ohta, J. S. Fritz, S. Matsushita and A. Miyanaga, Simultaneous ion-exclusion chromatography-cation exchange chromatography with conductimetric detection of anions and cations in acid rain waters, *J. Chromatogr. A*, 671, 239, 1994.

9 Special Techniques

9.1 Preconcentration

Ion chromatography is frequently used to determine anions and cations at very low concentration levels, often in the low μg/L (ppb) range. In the electric power industry the water used in steam generators must be almost free of Na^+, Cl^- and other ions to avoid stress corrosion cracking. The ionic content of ultrapure water used in the electronics industry must be kept to extremely low levels. Semiconductor chip manufacturers require clean-rooms with utility impurities of no more than 1 ppb for 0.35 μm devices [1].

As advances in technology have been made, IC detection limits have been lowered through careful control of the column and detector temperature and improvements in pump and detector design. One way to extend ion chromatography to further lower limits of detection is to use a larger volume of a liquid sample. A sample loop of 100 μL or even 200 μL could be used instead of a more typical 10 μL loop. By injecting 10–20 times more sample, the limits of detection for sample ions should also be lowered. However, the zone occupied by the sample solution in the IC column will also be larger. Since the sample ionic strength is much lower than that of the eluent there may be a prolonged dip in the chromatographic baseline when this sample zone passes through the detector.

A separate concentrator column is most commonly used to extend the working range of ion chromatography to significantly lower levels. A concentrator column is a short column (typically 35–50 mm in length) placed in a valve just before the analytical column. Sometimes a guard column is used as the concentrator column. The concentrator contains a stationary phase that is identical or similar to the analytical separator column. The function of a concentrator column is to strip ions from a relatively large volume of an aqueous sample of very low ionic content. A valving arrangement enables the sample to pass through the concentrator column directly to waste. Then a valve is switched and the ions taken up by the concentrator column are swept into the analytical column by the eluent stream where they are separated chromatographically. The advantage of this system is the ability to perform routine analyses for ions at low μg/L (ppb) levels.

Although several valve arrangements may be used, the basic configuration is illustrated by Fig. 9.1. In the *load* mode the sample flows through the concentrator column

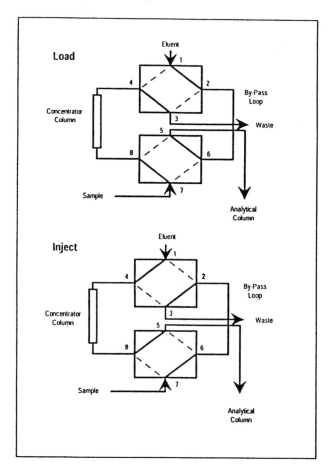

Figure 9.1. Configuration for a Dionex Low Pressure 4-Way Valve and a Concentrator Column (Courtesy Dionex Corp).

and out to waste (7, 8, 4, 3 sequence). Simultaneously, the eluent bypasses the concentrator column and flows into the analytical column (1, 2, 6, 5 sequence). In the *inject* mode the valve is switched so that the eluent flows through the concentrator column in the opposite direction to sample loading and into the analytical column (1, 4, 8, 5 sequence). Simultaneously, the sample stream is directed to waste (7, 6, 2, 3 sequence). Sample introduction may be by a small pump or with a manual syringe.

The sample breakthrough volume from the concentrator column needs to be measured in order to know how large a sample may be used. The sample must be of low ionic strength (<50 µS), otherwise the sample itself can act as an eluent for the sample ions. A good discussion of the use of concentrator columns in IC is available [2].

9.2 Sample Pretreatment

Samples that do not require preconcentration may contain substances that may adversely affect chromatographic performance. These substances may mask peaks of interest or be irreversibly retained, permanently damaging the column. In such cases sample pretreatment is necessary before an IC separation can be attempted.

9.2.1 Neutralization of Strongly Acidic or Basic Samples

Determination of small concentrations of anions in samples containing 1 M sodium hydroxide and determination of trace cations in strong mineral acids are two examples where sample neutralization is needed. However, chemical neutralization with HCl or NaOH would introduce a high concentration of unwanted ions. A better way is to introduce the H^+ or OH^- needed for neutralization through an ion-exchange membrane. This process is called electrodialysis.

Haddad, Laksana and Simons [3] described a device for off-line neutralization of strongly alkaline samples. The method uses an electrodialysis cell comprising three compartments separated from each other by cation-exchange membranes (Fig. 9.2).

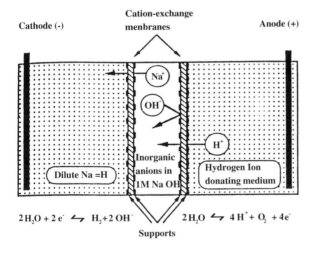

Figure 9.2. Schematic diagram of the electrodialysis process (from Ref. [3] with permission).

Cations can pass through the membrane but anions cannot. The cathode compartment of the cell was typically filled with 20 mL of 0.1 M sodium hydroxide, the anode compartment with 10 mL of 0.05 M sulfuric acid, and the sample compartment contained 2 mL of highly alkaline sample. Application of power (< 3 W) resulted in currents of 100–200 mA. As shown in the figure, water is reduced to H_2 + OH^- in the cathode and is oxidized to O_2 + H^+ in the anode compartment. To complete the electric circuit cations must be carried from the anode through the sample solution to the cathode.

Hydrogen ions entering the sample compartment will be neutralized by OH⁻ to form water, and Na^+ from the sample compartment will move across the membrane into the cathode compartment. The net result is that both Na^+ and OH⁻ are removed from the sample without disturbing other sample anions.

Devices are commercially available that maintain sample integrity and trace detection limits, while eliminating interfering sample matrices. Dionex offers an Auto Neutralization Module that neutralizes concentrated acid with electrically generated hydroxide ions [4].

Workers at Alltech [5] described two on-line methods for sample preparation. An ERIS Autosuppressor removes high concentrations of metal ions in samples to be analyzed for organic acids. It contains a cation exchanger that retains metal ions while allowing organic acids to pass through the cell. The ion exchanger is electrically regenerated. The device is constructed so that one cell is regenerated while the other cell is in use. Then the cells are reversed for the next run. Another device, the Alltech SCAN Sample Processor, is designed both for sample concentration and neutralization. The cell is packed with either anion or cation-exchange resin, depending on the application.

9.2.2 Particulate Matter

The sample solutions injected into an ion chromatographic system must be free of particulate matter to avoid plugging of the capillary connecting tubing and the frits at the head of the analytical column. Even samples that appear to be clear may contain unsuspected fine particles. It is more or less standard procedure to filter sample solutions prior to their injection. Disposable membrane filters with a pore diameter of 0.45 µm are sufficient in most cases. Samples with biological activity are filtered through asceptic filters with a pore diameter of 0.22 µm to avoid a change in sample composition due to bacterial oxidation or reduction. In general, membrane filters should be rinsed with de-ionized water prior to use to avoid sample contamination.

9.2.3 Organic Matter

Solid or semi-solid samples may require extraction with an aqueous solution to isolate the ionic components in a form suitable for IC. The actual procedures vary widely, depending on the type of sample. For example, meat and sausage products to be analyzed for nitrate and nitrite are first homogenized mechanically, extracted with a 5 % borax buffer solution in a hot water bath, and then subjected to a precipitation with strong solutions of potassium hexacyano ferrate and zinc sulfate. The aqueous extracts are diluted further with de-ionized water and filtered through a membrane prior to injection.

Organic substances in the sample matrix may interfere with ion chromatographic separations. In some cases it is sufficient to add enough methanol or another organic solvent to completely solubilize organic matter. But in samples that are soluble in water alone, the organic components may be taken up by the IC column packing and prevent reproducible results. This often occurs with dyes that are added to many commercial products.

Dyes and many other types of soluble organic compounds can usually be removed by some form of solid-phase extraction (SPE) without altering the inorganic ion content of the sample. In simple cases a SPE cartridge or membrane disk is attached to the sample syringe. Then the liquid sample is injected through the SPE material into the ion chromatograph. SPE cartridges attach to the sample syringe by a Luer tip. A solid disk can be cut from a larger SPE disk, such as the Empore disks produced by the 3M Co., and fitted tightly inside the sample syringe. A convenient semi-micro device for SPE with a syringe has been described [6].

Many methods are available for removal of organic material from aqueous samples by off-line SPE [7]. Hydrophobic organic material is best extracted by solid poly(styrene–DVB) polymers or reversed-phase silica extractants. Polyvinylpyrrolidone (PVP) is an appropriate choice for removal of humic acids, lignins and tannins from water samples.

9.2.4 Dialysis Sample Preparation

Sample preparation by dialysis is where sample ions transfer or migrate across a membrane from its original matrix into a receiving solution. Then, the receiving solution containing the sample ions is injected into the IC and the ion analysis is performed. An attractive feature of the dialysis process is that it isolates the sample matrix from the sample ions that are injected onto the analytical column. Some applications of dialysis sample pretreatment include the determination of ions in milk and dairy products or canned goods after being homogenized. Fruit juices containing pulp can be determined for ionic content.

Depending on the type of membrane used, dialysis may be either passive or active (Donnan type). Passive dialysis employs a neutral porous membrane with a fine pore structure. This type of dialysis is used primary to remove sample matrix materials such as proteins and/or fiber or other solids [8,9]. Donnan dialysis can do the same, but can also perform preconcentration of the ions in sample. In Donnan dialysis, an ion exchange membrane is used to exchange sample ions for receiving solution ions [10–13]. Each type of dialysis is described below.

Passive Dialysis: This type of dialysis provides an attractive method to remove any undesired, solid or suspended material from a IC sample that would normally be difficult to filter. The technology is based on a hydrophilic membrane containing small pores through which ions and other small neutral material can migrate, but larger particles cannot pass. A typical membrane is a hydrophilic polysulfone ultrafilter designed to filter materials in the middle molecular weight range of 500–2000. The membrane may be hollow fiber or sheet type. A hollow fiber membrane is used where the sample is pumped into the hollow fiber and ions migrate out of the fiber into the receiving solution located on the outside. Generally, both the sample and receiving solution are nonflowing or static. Older dialysis procedures based on static sample and receiving solutions have been slow and have required relatively large samples. These procedures have usually resulted in severe dilution of the sample.

Newer dialysis methods are based on a flowing sample stream and a static receiving solvent (usually water) [9]. In this type of apparatus, a sheet membrane is used to separate sample and receiving channels. A typical static receiving channel volume is 240 µL. The flowing sample stream channel has a similar volume, but the sample is flowing with a typical flow rate 0.8 mL/min. The sample continues to flow until the ionic concentration comes to an equilibrium. After dialysis, typically 20 µL of the receiving solution is injected into the IC. Using this method, 100 % of the sample ionic concentration can be transferred to the acceptor solution. The sample solution will come to equilibrium with the receiving side so that equal concentrations of ions are on both sides of the membrane. Calibration can be carried out easily with external standards or, if desired, internal standards.

The sample must only be in a liquid or suspension in a homogeneous form before introduced into the dialyzer. Heavy solids are separated from the sample by centrifugation before introduction of the fluid into the dialyzer. The time required for the dialysis is about the same as the IC run time so the dialysis time can be overlapped with elution of the previous sample. This type of system lends itself to fully automated sample handling.

Donnan (active) Dialysis: Donnan dialysis provides an alternative to column ion-exchange, filtration, or precipitation methods that are used either to preconcentrate ions and/or remove interfering ions. In Donnan dialysis, an (aqueous) sample is separated from a receiving electrolyte by an ion exchange membrane. Only the desired ions and a small amount of neutral molecules go form the sample into the receiver. Depending on the charge of the membrane, either anions or cations can travel across the membrane. An anion exchange membrane with fixed positive charges, will allow the transport of anions across the membrane, but will not allow the transport of cations. The reverse is true for a cation exchange membrane. Transport of cations is allowed and transport is not allowed for anions.

There are several processes that are occurring in Donnan dialysis sample preparation. One process involves the normal transport ions of like charge across the membrane. At the start of dialysis there is a strong tendency for the any particular ion to diffuse from a high concentration zone on one side of the membrane to the low concentration zone on the other side. As this process occurs, corresponding transfer of the same charge ion travels back across the membrane in order to preserve electro-neutrality. The counter ion does not travel across so there can exist different normality concentrations of ions on each side. The example of 90 % transport of a sample (Na^+) Cl^- at 0.0010 N and a receiving electrolyte (Na^+) OH^- at 0.050 N at 1 mL each is shown in Fig. 9.3.

Since, NaOH is at the higher concentration, the driving force is to make the ionic concentrations equal on both sides. But Na^+ is prevented from traveling across the membrane. Instead, most of the Cl^- is replaced by OH^-. The Cl^- is never completely recovered on the receiver side, but most will travel across because the OH^- is at a much higher concentration. Since OH^- is at a much higher concentration, its dilution by transport to the sample is not greatly affected.

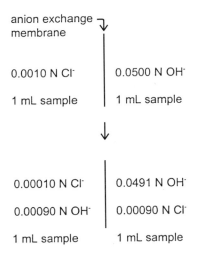

Figure 9.3. Example, with Cl⁻ dialyzed into a 0.0500 N OH⁻ receiving solution where the sample and receiving volume are equal. In this example, 90 % of the sample Cl⁻ enters the receiving solution and a corresponding amount of OH⁻ replaces it in the original sample solution. The counterions such as Na^+ or K^+ do not transport across the membrane. The exact amount of Cl⁻ that transports across is time-dependent.

The exact amount of sample ion transported across is not necessarily 90 %. The amount is affected by the contact time, the mixing of the solutions and probably other factors. That is why standards used to calibrate the instrument should undergo the same process as the sample. Internal standard spikes added to the sample are the best way to ensure proper calibration.

Now, consider when the volume of the sample is 10 times greater than the receiver. There is ten times more Cl⁻ available for transport and in this example 90 % of the chloride is transported. The receiver concentration of the OH is reduced to a greater extent, but the benefit is the Cl⁻ transported across is almost 10 times the original concentration (see Fig. 9.4).

Finally, consider the case where the sample is flowing past the membrane and the receiver remains static. In this case, only the volume and concentration of the sample and the concentration of the receiver solution limit the concentration effect of the sample into the receiver solution. Obviously, sample anions can only be transported across the membrane if there are receiving ions available for transport back across. Electric neutrality must be preserved.

This last phenomenum is important to understand. It is to the benefit to the Donnan dialysis process to have a high receiving ion concentration. But depending on the solution type and concentration, subsequent injection of this solution (that now contains the sample) can interfere with the ion chromatography analysis. Ideally, the receiving solution type should be the same as the eluent used for the IC analysis. A sodium carbonate/bicarbonate receiving solution is used commonly for suppressed IC. Also, it is possible to remove the receiving solution background ions prior to injection. This is accomplished by treating the receiving solution containing the sample anions with a H-form column or membrane. This converts all of the anions to the acid form

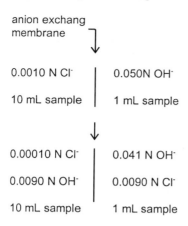

Figure 9.4. Example, with Cl⁻ dialyzed into a 0.050 N OH⁻ receiving solution where the sample volume is ten times greater than the receiving volume. In this example, 90 % of the sample Cl⁻ also enters the receiving solution and a corresponding amount of OH⁻ replaces it in the original sample solution. But because of the volume differences, the concentration of Cl⁻ is increased in the receiving solution relative to the sample solution.

and effectively removes the carbonate/bicarbonate through conversion to carbon dioxide. The process gives high enrichment factors, and wide linear dynamic range, and is free of many sample matrix effects.

Electrochemical dialysis is a refinement to the dialysis method (discussed in Section 9.2.1). The transfer of ions through a membrane is driven by the application of an electric field.

9.2.5 Isolation of Organic Ions

Organic compounds that are ionic or that become ionic at certain pH values can be isolated selectively from neutral organic compounds by SPE with ion-exchange materials [7]. Amines become protonated cations in acidic solution and are retained by a short cation-exchange column in the H⁺ form. Actually, the cation exchanger or exchanges can convert a neutral amine to the protonated cation.

$$RNH_2 + catex\text{–}H^+ \rightarrow Catex\text{–}RNH_3^+$$

Neutral organic compounds that cannot exist as cations may be retained by physical adsorption but can be washed off the cation exchange column by a brief rinse with an organic solvent. The amine cation can then be eluted from the column with a 1 M solution of trimethylamine in methanol. The trimethylamine converts the amine cation to the free amine which is no longer retained by the cation exchanger. Because of its volatility, trimethylamine is easily removed from the eluate. After acidification, the sample amines can be separated by cation chromatography.

The cation exchange resin used in this operation should be a macroporous, rather than a microporous, polymer to minimize physical adsorption. An intermediate exchange capacity (~0.6–1.2 mequiv/g) is required and the column dimensions should be small to avoid use of an excessive amount of trimethylamine–methanol solution to neutralize the resins H$^+$ capacity in the elution step. An Empore ion-exchange disk inserted into a tube is ideal for this use.

In a similar manner, carboxylic acid anions, phenolates, etc. can be separated from other organic matter by retention on a small anion exchanger in the OH$^-$ form.

$$RCO_2H + Anex–OH^- \rightarrow Anex–RCO_2^- + H_2O$$

After a brief rinse the retained sample anions are eluted with a solution of 1 M HCl in methanol. Again, the ion exchange column should be small and the resin should be a macroporous polymer of ~1 mequiv/g exchange capacity.

9.3 Ion-Pair Chromatography

9.3.1 Principles

As an alternative to conventional chromatography, an inorganic or organic anions (A$^-$) or cation (C$^+$) may be separated by adding a large organic ion (O$^+$ or O$^-$) of the opposite charge to the mobile phase. Presumably, this would result in the formation of an ion pair (O$^+$A$^-$ or O$^-$C$^+$) that was more "organic" in nature than the sample ion and could therefore be separated on an ordinary HPLC column with an organic–aqueous mobile phase such as methanol–water or acetonitrile–water. This type of separation does indeed work well in many cases provided the mobile phase also contains the ion pairing reagent. A typical ion pairing reagent for sample cations would be a quaternary ammonium salt such as a tetrabutylammonium salt. Sodium salts of alkanesulfonic acids have been used as ion-pairing reagents for sample cations.

The mechanism of what we will call "ion-pair chromatography" has been the subject of a considerable amount of investigation. Horvath et al. demonstrated the practicality of this approach [14]. They proposed an ion-pair mechanism and gave a number of ion-pair formation constants.

Kraak, Jonker and Huber used anionic surfactants in conjunction with a bonded-phase silica column and an organic–aqueous mobile phase for the separation of amino acids [15]. A comprehensive study was made of the parameters, including the generation of gradients.

Kissinger argued that an ion-pair mechanism is incorrect [16]. The pairing reagent partitions strongly onto the stationary phase, modifying its surface charge. This implies an ion-exchange mechanism. This interpretation would appear to be valid when the pairing reagent is very strongly adsorbed on the stationary phase surface. This situation is much like using a reversed-phase HPLC column packing with a permanent coating of a surfactant for ion chromatography.

In a more common form of "ion-pair" chromatography a smaller ion pairing reagent is added to the organic-aqueous mobile phase. This mobile phase is pumped through a reversed phase HPLC column until equilibrium is attained. Then a sample is injected and the sample ions opposite in charge to the ion-pairing reagent are separated. Bidlingmeyer et al. have discussed the mechanism for this type of chromatography [17]. Suppose that sample cations are to be separated with sodium octanesulfonate added to the eluent as the ion-pairing reagent. As the column is "conditioned" by pumping through the mobile phase, octanesulfonate is adsorbed creating a negatively-charged primary layer on the reversed-phase surface. The positively charged sodium ions form a second layer. Because this equilibrium is dynamic (octanesulfonate ions are continually desorbed and others adsorbed), negative solute ions can compete and effect a net displacement. This can give rise to a displacement peak. The secondary layer is also a dynamic equilibrium so that positively charged sample ions may exchange with sodium ions. The net result is that a pair of ions (octanesulfonate and sample ions, C^+), but not necessarily an ion pair, has been adsorbed onto the stationary phase. The name "ion-interaction chromatography" has been applied to this type of separation.

9.3.2 Typical Separations

A variety of inorganic and organic ions (both anions and cations) have been separated by this type of chromatography. Haddad and Jackson give numerous examples [18]. The following experimental parameters can be adjusted to obtain satisfactory conditions for a separation:
– Type of stationary phase
– Type of pairing reagent. A higher molecular weight shifts the equilibrium towards the surface.
– Concentration of pairing reagent. A higher concentration also shifts the equilibrium towards the surface.
– Type and concentration of organic solvent (organic modifier). Increasing concentrations shift the equilibrium away from the surface.

The ionic pairing reagent necessarily introduces a counterion into the system. This ion preferably should be different from any of the sample ions to be determined. Some of the most useful separations are of organic ions or ionic inorganic complexes. The separation of eight metal cyanide complexes in Fig. 9.5 would be difficult to accomplish by conventional ion chromatography. Detection in this case was by direct spectrophotometry at 214 nm.

Suppressed conductivity detection may be used in some cases. A micromembrane suppressor is available that is solvent resistant and is permeable to quaternary ammonium ions. A tetrabutylammonium hydroxide eluent in aqueous–organic solution is often used in the separation of anions. The tetrabutylammonium ion is sufficiently large to act as a good ion-pairing reagent. In the suppressor unit the OH^- counter ion is converted to water while the counter ion of sample anions is converted to the highly conducting H^+. A separation of C5–C8 alkylsulfonates with suppressed conductivity detection is shown in Fig. 9.6. The mobile phase consisted of 2.0 mM tetrabutylammonium hydroxide in 37 % acetonitrile [19].

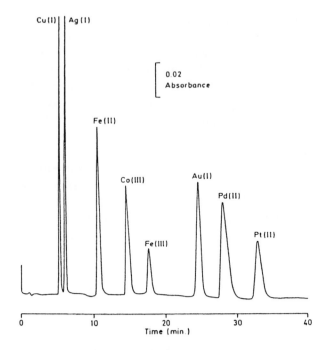

Figure 9.5. Ion-interaction separation of metal–cyano complexes. A Waters Nova Pak C_{18} column was used with 23:77 acetonitrile–water containing 5 mM Waters Low UV PIC A as eluent. Detection was by direct spectrophotometry at 214 nm (from Ref. [18] with permission).

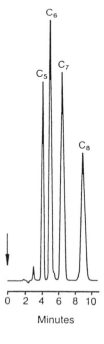

Figure 9.6. Separation of alkylsulfonates by ion-pair chromatography. Eluent: 2.0 mM tetrabutylammonium hydroxide in 37 % acetonitrile, suppressed conductivity detection. Source: Ref. [19].

9.4 Simultaneous Separation of Anions and Cations

A dream of ion chromatographers is to separate both anions and cations in a single run. In 1986, Pietrzyk and Brown used a mixed bed of anion- and cation-exchange resins to separate a mixture of fluoride, chloride, sodium, potassium and nitrate in ~20 min [20]. A mobile phase containing lithium acetate was used in conjunction with conductivity detection. Later, anion–cation separations were performed with indirect photometric detection. Thus far, only moderate separation efficiency has been achieved [21].

A group in Japan has pioneered the simultaneous separation of anions and cations using a column packing that is coated with a hydrophobic zwitterionic reagent [22]. Their method has the unique advantage of using only pure water as the mobile phase. This provides an unusually low background signal with either conductivity or UV–visible detectors that is ideal for the separation and detection of very low concentrations of ionic sample components.

When a small volume of an aqueous sample containing anions and cations is passed through a column with a zwitterionic stationary phase, neither the cations nor anions can get very close to the opposite charge on the stationary phase. The sample anions and cations are forced into a new state of simultaneous electrostatic attraction and repulsion interaction. Thus the sample ions are somewhat attracted to the zwitterionic stationary phase, but the attraction is weak enough that water alone can serve as the eluent. To preserve electroneutrality, an equal charge of anions and cations must be eluted. The authors have termed this separation method as EKC (electrostatic chromatography).

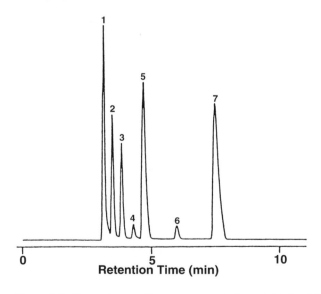

Figure 9.7. Chromatogram of an aqueous solution containing 2.857 mM each of Na_2SO_4, NaCl, NaBr, NaI, NaSCN, CaI_2, and $Ca(SCN)_2$. Separation conditions and detection (conductivity) are the same as in Fig. 9.4. Key: (1) 2 Na^+–SO_4^{2-}, (2) Na^+–Cl^- (3) Na^+–Br^-, (4) Ca^{2+}–$2Br^-$, (5) Na^+–I^-, (6) Ca^{2+}–$2I^-$, (7) Ca^{2+}–$2SCN^-$ (From Ref. [22] with permission).

A complicating factor in EKC is that peaks are observed corresponding to different combinations of anions and cations. Figure 9.7 shows the separation of several ions using a column packing coated with CHAPS, 3-[3-cholamidopropyl)dimethylamino]-1-propane sulfonate. Although ten peaks were predicted, the peaks corresponding to $Ca^{2+}SO_4^{2+}$, $Ca^{2+}2Cl^-$ and $Na^{2+}SCN^-$ were missing. There is a certain priority that governs which ion pairs will form.

In another paper, the complication of multiple peaks in EKC was confirmed [23]. The use of a preliminary ion exchange column was suggested to give only a single common cation or anion. This method preserves the use of pure water as the eluent but takes away the ability to separate mixtures of anions and cations simultaneously.

A method for determination of cations and anions in a single run was described in Section 8.6. The column was packed with a polyacrylate gel cation exchanger. It appears that sample cations were separated by cation chromatography and anions by ion-exclusion chromatography.

References

[1]   S. J. Lue, T. Wu, H. Hsu and C. Huang, Application of ion chromatography to the semiconductor industry I. Measure of acidic airborne contaminates in clean rooms, *J. Chromatogr. A*, 804, 273, 1998.

[2]   Dionex Corp. (Sunnyvale, CA U.S.A.), the use of concentrator columns in ion chromatography. 1994.

[3]   P. R. Haddad, S. Laksana and R. G. Simons, Electrodialysis for clean-up of strongly alkaline samples in ion chromatography. *J. Chromatogr.*, 640, 135, 1993.

[4]   A. Siriraks and J. Stillian, Determination of anions and cations in concentrated bases and acids by ion chromatography, *J. Chromatogr.*, 640, 151, 1993.

[5]   R. M. Montgomery, R. Saari-Nordhaus, L. M. Nair and J. W. Anderson, Jr., On-line sample preparation techniques for ion chromatography, *J. Chromatogr. A*, 804, 55, 1998.

[6]   D. L. Mayer and J. S. Fritz, Semi-micro solid-phase extraction of organic compounds from aqueous and biological samples, *J. Chromatogr. A*, 773, 189, 1997.

[7]   J. S. Fritz, Analytical solid-phase extraction. Wiley–VCH, New York, 1999.

[8]   F. R. Nordmeyer and L. D. Hansen, Automatic dialyzing-injecting system for liquid chromatography of ions and small molecules, *Anal. Chem.*, 54, 2605, 1982.

[9]   Metrohm literature, On-line sample preparation in ion chromatography – no problem with the novel 754 dialysis unit, Metrohm Information, 74th ed., 26, 3, 1999.

[10] J. E. Dinunzio and M. Jubara, Donnan dialysis pre-concentration for ion chromatography, *Anal. Chem.*, 55, 1013, 1983.

[11] J. A. Cox and J. Tanaka, Donnan dialysis preconcentrator for the ion chromatography of anions, *Anal. Chem.*, 57, 2370, 1985.

[12] J. A. Cox and E. Dabekzlotorzynska, Determination of anions in poly-electrolyte solutions by ion chromatography after Donnan dialysis sampling, *Anal. Chem.*, 59, 543, 1987.

[13] S. Laksana and P. R. Haddad, Dialytic clean-up of alkaline samples prior to ion chromatographic analysis, *J. Chromatogr.*, 602, 57, 1992.

[14] C. Horvath, W. Melander, I. Molnar and P. Molnar, Enhancement of retention by ion-pair formation in liquid chromatography with nonpolar stationary phases, *Anal. Chem.*, 49, 2295, 1977.

[15] J. C. Kraak, K. M. Jonker and J. F. K. Huber, Solvent generated ion-exchange systems with anionic surfactants for rapid separation of amino acids, *J. Chromatogr.*, 142, 671, 1977.

[16] P. T. Kissinger, Comments on reverse-phase ion-pair partition chromatography, *Anal. Chem.*, 48, 883, 1977.

[17] B. A. Bidlingmeyer, S. N. Deming, W. P. Price, J. B. Sachok and M. Petrusek, Retention mechanism for reversed phase ion-pair liquid chromatography, *J. Chromatogr.*, 186, 419, 1979.

[18] P. R. Haddad and P. E. Jackson, Ion Chromatography, Principles and Applications, p 189. Elsevier, Amsterdam, 1990.

[19] J. Weiss, Ion Chromatography, 2nd ed., p 251. VCH, Weinheim, Germany, 1995.

[20] D. J. Pietrzyk and D. M. Brown, Simultaneous separation of inorganic anions and cations on a mixed-bed ion exchange column, *Anal. Chem.*, 58, 2554, 1986.

[21] D. J. Pietrzyk, S. M. Senne and D. M. Brown, Anion–cation separations on a mixed bed ion-exchange column with indirect photometric detection, *J. Chromatogr.*, 546, 101, 1991.

[22] W. Hu, H. Tao and H. Haraguchi, Electrostastic ion chromatography, 2. Partition behaviors of analyte cations and anions, *Anal. Chem.*, 66, 25414, 1994.

[23] T. Umemura, R. Kitaguchi and H. Haraguchi, Counterionic detection by ICP-AES for determination of inorganic anions in water elution ion chromatography using a zwitterionic stationary phase, *Anal. Chem.*, 70, 936, 1998.

10 Capillary Electrophoresis

10.1 Introduction

Capillary electrophoresis (CE) is a method in which ions are separated by differences in their rates of migration through a silica capillary. The capillary is filled with an electrolyte solution and each end of the capillary dips into an electrolyte reservoir which also contains a platinum electrolyte. The electrodes at the two ends of the capillary are connected to a high-voltage power supply (0–30 kV). Ions in solution will flow through the capillary to complete an electric circuit. The sample ions to be determined migrate at different velocities towards the electrode of opposite charge (electrophoresis flow). The sample ions are detected spectrophotometrically as they pass through a cell near the end of the capillary.

Separations by ion chromatography and capillary electrophoresis are both based on differences in the velocities at which ions move through a column or capillary. However, in IC these differences in velocity are the result of differences in partitioning of sample ions between a stationary ion exchanger and the liquid mobile phase. In CE there is no partitioning between the two phases; differences in velocity are the result of differences in electrical mobility (electrophoretic mobility) through an non-packed capillary.

CE has several advantages over IC. In terms of theoretical plates, CE has at least ten times the separation of a typical IC system. Separations by CE are fast and it is relatively easy to adjust experimental conditions to obtain an adequate separation of sample ions. CE is a truly micro method that permits the use of very small samples. However, CE has a number of drawbacks that include a limited choice of detectors, a mediocre detection sensitivity, occasional problems with reproducibility, and a perception that CE is a somewhat exotic technique that is more complicated than IC.

The purpose of this chapter is to present a compact treatment of the principles of CE and an idea of its scope as applied to the analysis of inorganic and small organic ions.

10.1.1 Experimental Setup

The essential parts of a CE instrument are shown in Fig. 10.1. A background electrolyte (BGE) is placed in reservoirs (A) and then pumped through the capillary (C) to fill it with BGE. A typical capillary is made of fused silica, 50 or 75 µm i.d. and approximately 60 cm long, although the length may vary. A platinum working electrode (B) is placed in each reservoir and connected to a high-voltage power supply (E) capable of generating up to 30 kV. A positive power supply makes the left-hand electrode the anode (positive charge) and the other electrode the cathode. With a negative power supply these polarities are reversed. Conditions must always be arranged so that sample anions will migrate from left to right and thus flow through the detector.

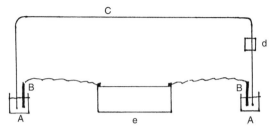

Figure 10.1. Schematic representation of a capillary electrophoresis instrument. A, electrolyte reservoirs; B, platinum electrodes; C, fused silica capillary; D, detector cell; E, high-voltage power supply.

10.1.2 Principles

Separations of anions are based on differences in electrophoretic flow. Inorganic ions are generally smaller and therefore more mobile than organic ions. The electrophoretic mobilities of inorganic ions are an inverse function of their hydrated ionic radii. Electrophoretic mobility is also affected by the charge on an ion and by the solvent medium. Tables of limiting ionic conductance are a convenient source for estimating electric mobilities of ions.

In addition to electrophoretic flow, a second type of migration occurs in the capillary called electroosmotic flow (EOF). A minute charge on the capillary surface (called the zeta potential) results from ionization of silanol groups, which have a pK_a of 6 to 7.

$$-SiOH \rightleftarrows SiO^- + H^+$$

Cations from the bulk solution are attracted to the capillary surface. These cations are attracted toward the cathode. Since the cations are hydrated and the inner diameter of the capillary is quite small, their migration induces a bulk liquid flow (a plug flow). This means that all of the liquid and solutes in the capillary flow at the same rate towards the cathode (electroosmotic flow). The mobility of solute resulting from EOF is termed the electroosmotic mobility (μ_{os}).

EOF in fused silica capillaries is pH-dependent. The EOF is very low around pH 3–4, but rises rapidly around pH 5–6 as the silanol groups become progressively more ionized. Finally, the EOF levels out around pH 8–9.

The magnitude and direction of electromigration can be indicated by vectors. For sample cations in neutral to alkaline solution, the electrophoretic and electroosmotic flow vectors are in the same direction. This condition is called comigration. The electrophoretic vectors of anions are in the opposite direction to the cationic vectors. Usually the electroosmotic vector is larger than the electrophoretic vector, so the net movement of anions is still toward the cathode. This is important because the separated sample ions (anions in this case) must all pass through the fixed detector. A separation in which the electrophoretic and electroosmotic vectors are in the opposite direction is called counter migration. A separation with counter migration will take longer than one with comigration, but the separation power of counter migration is usually much better.

10.1.3 Steps in Analysis

Capillary pretreatment: A separation with sharp peaks for sample ions and reproducible migration times requires a clean capillary surface. This is usually accomplished by frequent rinsing of the capillary with dilute aqueous sodium hydroxide. After a water rinse, the capillary is filled with the BGE solution. The BGE contains a pH buffer and a sufficient concentration of an electrolyte to maintain a steady current (frequently 20–50 mM).

Sample introduction: To introduce the sample, the left end of the capillary is dipped into a sample vial with several centimeters hydrostatic pressure for a fixed number of seconds. This will force a small volume of liquid sample into the end of the capillary. Another method of sample introduction is to dip the end of the capillary into the sample vial and turn on the power for a few seconds. Sample ions thus flow into the system by electrical migration.

Sample run: When the sample has been added to the system the power is turned on. Most commonly a power of + or –20, 25 or 30 kV is applied, although sensitive samples may necessitate a lower power. Sample ions will now migrate by electrophoretic flow toward the electrode of opposite charge (provided the electrophoretic flow is either the same direction or larger than the electroosmotic flow). Separations are based on differences in electrophoretic mobility. Tables of limiting equivalent conductivity are a good guide to predicting the relative rates of migration of sample ions. To separate several cations, a positive power supply would be used and the sample ions will migrate at different rates toward the detector end (negative electrode) of the system.

Many separations of anions require that the direction of EOF be reversed. This is accomplished by adding a flow modifier, such as a quaternary ammonium salt with a long hydrocarbon chain, to the BGE. A thin layer of the flow modifier is adsorbed on the capillary surface. This gives the surface a positive charge and causes electrolyte anions to give an electroosmotic flow towards the anode.

Detection: Sample ions absorb sufficiently in the UV or visible spectral region may be detected by direct spectrophotometry. Indirect spectrophotometric detection is commonly used for ions that do not absorb. An absorptive reagent is added to the

BGE gives a peak in the direction of reduced absorbance when a sample ion passes through the detector. The absorbing reagent, which is sometimes called a "visualization" reagent, should have a mobility that matches those of the sample ions as closely as possible. Chromate is often used for the indirect detection of anions and a protonated amine cation, such as benzylamine, for detection of cations.

10.2 Some Fundamental Equations

Table 10.1 lists several equations that apply to CE. Equation 1 states that the velocity (cm/s) with which an ion moves through the capillary is a function of its mobility and field strength. Field strength (V/cm) is defined in Eq. 2. Ionic mobility (cm^2/V s) is defined in terms of column length, migration time and applied voltage in Eq. 3. Equation 4 states that ionic mobility (cm^2/V s) is made up of electrophoretic and electroosmotic mobility. In CE as in chromatography the separation power is often stated by the plate number, N (also called the number of theoretical plates).

Table 10.1. Some fundamental CE equations.

1. Velocity	$V = \mu E$	μ = mobility, cm^2/V s
		E = field strength, V/cm
2. Field strength	$E = V/L$	V = applied voltage, V
		L = capillary length, cm
3. Mobility	$\mu = L_d L/tV$	L_d = length to detector, cm
		L = length (total), cm
		t = migration time, s
		V = applied voltage, V
4. Mobility	$\mu = \mu_{ph} + \mu_{os}$	μ_{ph} = electrophoretic mobility
		μ_{os} = electroosmotic mobility

10.2.1 Peak Shape

Mikkers et al. [1] concluded that symmetrical peaks are obtained only when the mobility of the carrier co-ion closely matches that of the analyte ion. If mobility of the analyte ion is higher, fronted peaks will result.

Hjerten studied conductivity differences at boundary between analyte zone and carrier electrolyte.

$$\Delta k = C_B \left[\mu_B \left(\mu_A - \mu_B \right) \left(\mu_R - \mu_B \right) \right] \tag{10.1}$$

(where C_B = sample ion concentration, μ_B = sample ion mobility, μ_A = electrolyte co-ion mobility, μ_R = electrolyte counter asymmetry)

- Δk should be minimized to minimize peak asymmetry.
- A high concentration, C_B, changes conductivity within the sample zone. This leads to uneven migration within the zone and broad, triangular peaks.

10.2.2 Electrostacking

This is a technique for injection that produces an unusually sharp, concentrated band of analyte ions in the front section of the capillary. The dependence of migration velocity (v) on the electric field (E) is:

$$v = (\mu_i + \mu_{Eo})E \tag{10.2}$$

where: μ_i = individual ion mobility; μ_{Eo} = electroosmotic mobility.

According to the electrostacking condition, the sample zones in hydrostatic injections must have much lower ionic strength than the carrier electrolyte. When the separation voltage is applied, the low ionic strength of the sample zone creates a significantly higher resistance than that sustained by the electrolyte, which consequently produces a higher field strength. The increased field strength of the sample zone forces the analyte ions to migrate faster than the electrolyte ions, which are subjected to a lower field strength. The net effect of this differential migration rate is the accumulation of sample ions inside a very narrow zone at the sample–carrier electrolyte boundary. This precondition, or electrostacking, occurs before the migration of the analyte ion zone through the bulk of the carrier electrolyte solution.

To promote electrostacking, the ionic concentration of BGE should be significantly higher than that of the sample.

10.3 Separation of Anions

10.3.1 Principles

Inorganic anions and small organic anions are most often separated around pH 8.5 with a negative power supply. For a reasonably fast separation, the direction of electroosmotic flow must be reversed so that electrophoretic and electroosmotic flow will both be toward the anode. This is usually accomplished by adding a reagent to the BGE that will thinly coat the capillary surface, giving it a positive charge. Hydrated electrolyte anions will now move through the detector toward the anode, thus providing an EOF in the desired direction.

A typical flow modifier is a quaternary ammonium salt with three methyl groups and a longer hydrocarbon chain with 14 or 16 carbon atoms. For convenience, these salts will be denoted as Q^+. Addition of a 1 to 5 mM concentration of Q^+ to the BGE sets up a dynamic equilibrium between the liquid phase and the capillary wall. The positive end of the Q^+ molecule is probably attracted to the negatively charged silanol groups with the long hydrocarbon chain sticking out from the wall. Additional Q^+ molecules are hydrophobically attracted to these hydrocarbon tails with their N^+ end sticking out into the solution and away from the wall.

Some analyte anions absorb in the UV spectral region and can be detected by direct spectrophotometry. However, in many cases, indirect spectrophotometric detection is indicated. For indirect detection, a low concentration of an anion that absorbs strongly in the visible or UV regions is added to the BGE. This is called the "probe" ion or sometimes the "visualization reagent." A background signal is established by the probe ion passing through the detector at a fixed rate. Within a sample ion zone the concentration of the probe ion is reduced by an amount proportional to the sample ion concentration, thus resulting in a detection peak of reduced absorbance. The reason that the visualization ion concentration is lower within a sample zone is that the total ionic current in the capillary must remain constant.

A requirement of an anion used for indirect detection is that its mobility must match that of the sample ions as closely as possible. If the mobilities do not match reasonably well, peaks may be fronted or tailed. Since the mobilities of sample ions will differ, the mobility of the visualization reagent is matched as closely as possible to the mobility of the middle sample ions.

Chromate at a concentration of around 5 mM has been used very successfully for indirect detection of common inorganic anions at a wavelength of 254 nm. A now classic separation of some 30 anions in a single run is illustrated in Fig. 10.2. More recently, Shamsi and Danielson have proposed naphthalenedisulfonate (NDS) and naphthalenetrisulfonate (NTS) for indirect detection of inorganic and organic anions [2]. There is a good match in their migration times relative to those of common anions. Even though these are large anions, the migration times of NDS and NTS are short because of their 2– and 3– charges.

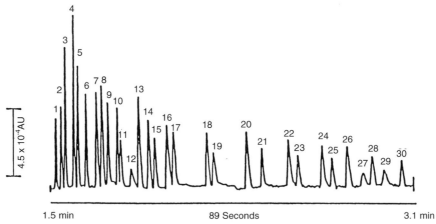

Figure 10.2. Peak identity and concentrations (ppm) for a 30-anion electropherogram displayed in an 89-s electropherographic segment. Electromigration injection at 1 kV for 15 s. Peaks: 1 = thiosulfate (4); 2 = bromide (4); 3 = chloride (2); 4 = sulfate (4); 5 = nitrite (4); 6 = nitrate (4); 7 = molybdate (10); 8 = azide (4); 9 = tungstate (10); 10 = monofluorophosphate (4); 11 = chlorate (4); 12 = citrate (2); 13 = fluoride (1); 14 = formate (2); 15 = phosphate (4); 16 = phosphite (4); 17 = chlorite (4); 18 = galactarate (5); 19 = carbonate (4); 20 = acetate (4); 21 = ethanesulfonate (4); 22 = propionate (5); 23 = propanesulfonate (4); 24 = butyrate (5); 25 = butanesulfonate (4); 26 = valerate (5); 27 = benzoate (4); 28 = L-glutamate (5); 29 = pentanesulfonate (4); 30 = d-gluconate (5). The electrolyte is a 5 mM chromate and 0.5 mM EOF modifier adjusted to pH 8.0 (From W. R. Jones and P. Jandik, *J. Chromatogr.*, 546, 445, 1991, with permission).

Comparison of the electrophoretic mobilities of several probe anions and analyte anions (Fig. 10.3) shows that chromate is a reasonably good match for fast-moving inorganic anions but is not so good for the other analytes listed. Lau showed convincingly that molybdate is a much better probe ion than chromate for indirect detection of common anions [3]. Sensitivity is better with molybdate molar absorptivity (5650 at 230 nm compared to 3180 at 254 nm for chromate), molybdate solutions are more stable, and peak shapes are better. Systematic studies resulted in the following optimal conditions: 5 mM molybdate as the UV-absorbing ion, 0.15 mM cetyltrimethylammonium hydroxide (CTAH) as an electroosmotic flow modifier, 0.01 % polyvinylalcohol as an additive to solve the comigration problem of fluoride and formate, and 5 mM tris(hydroxymethyl)aminomethane as a buffer to maintain a pH of 7.9. A separation of a standard anion mixture is shown in Fig. 10.4.

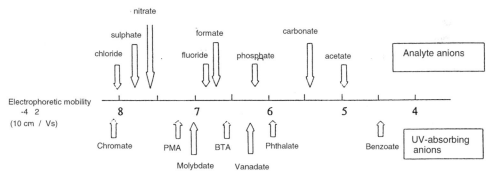

Figure 10.3. The electrophoretic mobilities of common UV-absorbing anions and analyte anions. From Ref. [3].

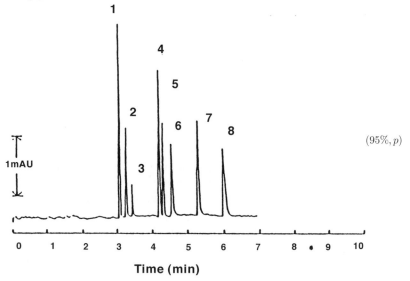

Figure 10.4. Separation of a standard anion mixture. BGE : 5 mM molybdate, 0.15 mM CTAH, 5 mM Tris buffer at pH = 7.9, 0.01 % PVA; Capillary: 65 cm × 0.075 mm i.d. fused silica; Run: –20 kV; Current: 12 μA; Injection: 8 cm for 20 s; Detection: 230 nm. Anions, 2 ppm each: 1 = chloride; 2 = sulfate; 3 = nitrate; 4 = fluoride; 5 = formate; 6 = phosphate; 7 = carbonate; 8 = acetate. From Ref. [3].

10.3.2 Separation of Isotopes

Separation of isotopes is an extremely challenging analytical problem. Lucy and McDonald [4] demonstrated the extraordinary separation power of CE by obtaining a baseline resolution of two chloride isotopes, $^{35}Cl^-$ and $^{37}Cl^-$. Counter migration was used and conditions were adjusted so that the electroosmotic and electrophoretic flow vectors were very similar in magnitude (but opposite in direction). Under these conditions the net migration velocity of the chloride ions is very slow but the separation power is very great. The following conditions were employed: –20 kV, pH 9.2 using a very low buffer concentration (3 mM borate), 6 mM chromate for indirect spectrophotometric detection (Fig. 10.5).

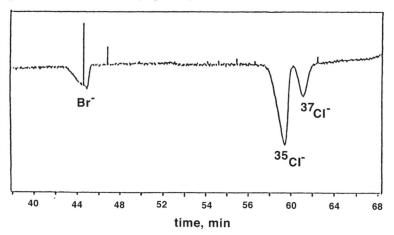

Figure 10.5. Separation of chloride isotopes by CE using counter migration (Courtesy of Youchun Shi).

10.3.3 Separations at Low pH

CE separations of inorganic anions are almost always carried out at an alkaline pH to ensure that the analytes will be in the ionic rather than the molecular form. However, a flow modifier must generally be used to reverse the direction of EOF. Thornton and Fritz found that excellent separations of anions are possible at pH values as low as 2.0 (HCl) or 1.8 with perchloric acid [5]. Special attention was paid to the anionic chloro complexes of gold (III) and the platinum group elements, which are more resistant to hydrolysis in more acidic solutions. By working at lower pH values, the capillary silanol groups are largely un-ionized and consequently the EOF is minimal.

Sharp peaks were obtained for gold(III) and for each of 24 species of chloro complexes of platinum group elements [5]. The theoretical plate number for the peak in Fig. 10.3 was 300,000. An applied potential of –10 kV was employed to reduce the high currents that occurred at –20 kV.

Sharp peaks were obtained for 14 other inorganic anions in hydrochloric acid or in perchloric acid (both pH 2.4) with direct detection at 214 nm. The migration times (Table 10.2) were very reproducible (average RSD = 0.6 %). The test anions included

MnO_4^-, VO_3^-, ReO_4^-, ferricyanide and ferrocyanide ions for which chromatographic separations are seldom possible.

Table 10.2. Migration times of anions at pH 2.4 in HCl–NaCl solution. Capillary 60 cm × 75 μm i.d. (52.8 cm to detector). Applied voltage –10 kV, direct detection at 214 nm.

Ion	t_M (min)	Rel. t_M ($Br^- = 1$)
Br^-	5.63	1.00
I^-	5.68	1.01
NO_3^-	6.26	1.11
$Fe(CN)_6^{3-}$	6.53	1.16
SCN^-	6.61	1.17
MnO_4^-	8.06	1.43
$HFe(CN)_6^{3-}$	8.33	1.48
IO_4^-	8.61	1.53
ReO_4^-	9.23	1.64
VO_3^-	9.61	1.71
$Cr_2O_7^{2-}$	10.06	1.79
IO_3^-	10.95	1.94
MoO_4^{2-} *	12.44	2.21
$HgCl_4^{2-}$	13.95	2.48

* pH 6.6

10.3.4 Capillary Electrophoresis at High Salt Concentration

It is commonly thought that even a moderately high ionic concentration in the BGE would lead to Joule heating and serious peak distortion. However, Ding, Thornton and Fritz found that very satisfactory separations of both inorganic and organic anions could be obtained in solutions as high as 5 M sodium chloride using direct photometric detection [6].

The first experiments on the effect of high salt concentrations were run at pH 2.4 to almost eliminate EOF. A negative power supply (–10 kV) and a 75 μm i.d. fused silica capillary were used. Both the sample and the BGE contained 0.5 M sodium chloride. The results for several inorganic anions under these conditions with direct photometric detection were poor. The peaks were badly shaped and there was almost no resolution of individual peaks. However, peak shape and resolution improved dramatically with increasing sodium chloride concentration in the BGE. At 1.5 M sodium chloride in the BGE, excellent separation was obtained for samples containing 0.5 M sodium chloride and low ppm concentrations of five inorganic anions. The salt content of the BGE needed to be at least three times that of the sample in order to provide sufficient peak focusing (electrostacking) during the sample introduction. These experiments demonstrated that the limits of salt concentration of both the sample and the BGE are much higher than had been expected.

A short study was conducted on the pH effect over a broad range, from pH 3.0 to pH 12.0. The sodium chloride concentration was 1.5 M in all buffers. There were no observed differences in either migration times or peak shapes for I^-, SCN^-, NO_3^- and IO_3^- between pH 3.0, 7.0, and 12.0. This effect strongly indicates that the electroosmotic flow is greatly suppressed, and the ionized silanol groups at the capillary surface are effectively shielded by the high concentration of cations, M^+, in the buffer solution [7].

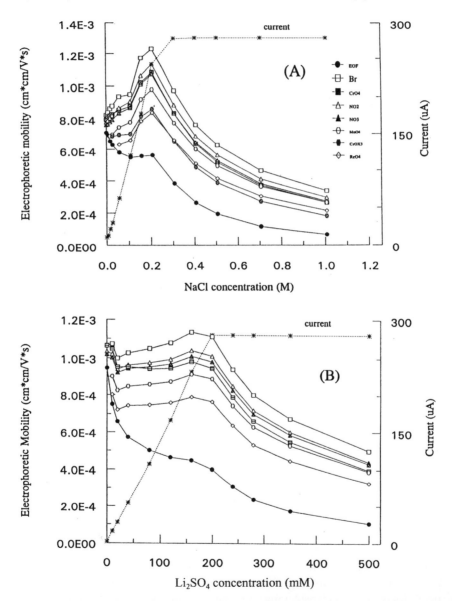

Figure 10.6. Plots of electrophoretic mobility and current against the concentration of the salts: (A) sodium chloride, and (B) lithium sulfate. Ions (top to bottom): Br^-, Cr_4^{2-}, NO_2^-, NO_3^-, MnO_4^-, $CrOx_3^{3-}$, ReO_4^-. From Ref. [6], with permission.

Sample solutions containing low concentrations of several inorganic anions were run at pH 8.5 with increasing concentrations of sodium chloride or lithium sulfate in the BGE. The plots in Fig. 10.6 show several interesting effects. One is that the current increases rapidly with increasing salt concentration and levels out at 280 μA around 200 mM sodium chloride or lithium sulfate. This sharp increase in current can be attributed to less electrical resistance. The maximum current that can be obtained in our instrument is set at 280 μA. In order to maintain this current, the voltage was automatically lowered as the salt concentration in the BGE continued to increase. The full power of the instrument's power supply was then being used.

The electrophoretic mobilities of the sample anions increased at the same time the current was increasing between 0 and 200 mM of added sodium lithium sulfate (Fig. 10.6). Actually, a decrease in electrophoretic mobility is predicted with increasing salt concentration. The initial increases can be explained by Joule heating. Under the conditions used a temperature of 49°C was calculated fore the capillary at high salt concentrations [6].

Figure 10.6 shows that the greatest differences between electrophoretic mobilities and electroosmotic mobility occur around 200 mM salt in the BGE. Figure 10.7 shows an excellent separation of inorganic anions at pH 8.5 in 220 mM sodium chloride. The high salt concentration suppresses the EOF sufficiently that no flow modifier is required, even at pH 8.5.

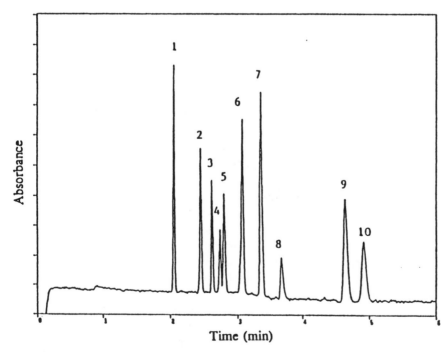

Figure 10.7. CE separation of ten inorganic anions. The 200 mM NaCl was added in the carrier electrolyte. Peaks: 1 = Br^- (10 ppm); 2 = NO_2^- (20 ppm); 3 = $S_2O_3^-$ (80 ppm); 4 = NO_3^- (2 ppm); 5 = N_3^- (40 ppm); 6 = $Fe(CN)_6^{4-}$; 7 = MnO_4^{2-} (40 ppm); 8 = WO_4^{2-} (40 ppm); 9 = $CrOx_3^{3-}$ (40 ppm); 10 = ReO_4^- (40 ppm). From Ref. [6], with permission.

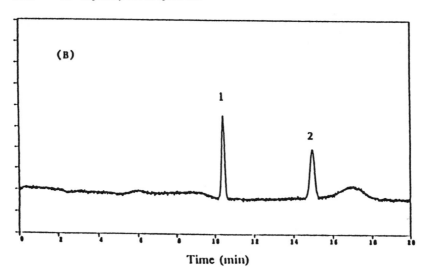

Figure 10.8. CE analysis of other high salt samples. Sample was in 0.5 M sodium sulfate, 1.5 M sodium sulfate was added to the electrolyte. Peaks: 1 = Br^- (10 ppm); 2 = NO_3^- (2 ppm). From Ref. [6], with permission.

CE of anions in solutions of high salt content has a number of practical applications. Bromide and nitrate in seawater were detected without any pretreatment or dilution of the samples. Anions in other high-salt samples can also be analyzed directly. Figure 10.8 shows peaks for 10 ppm bromide and 2 ppm nitrate in 0.50 M sodium sulfate. The BGE contained 1.5 pure sodium sulfate. It was also possible to determine both bromide and nitrate in 0.5 M sodium perchlorate by using 1.5 M sodium perchlorate in the BGE. With 1.5 M sodium chloride in the BGE, perchlorate and nitrate coeluted and only bromide could be measured.

10.4 Separation of Cations

10.4.1 Principles

Free metal cations are best separated in acidic solution in order to avoid problems due to hydrolysis. A positive power supply is generally used so that electrophoretic migration will be toward the cathodic detector end of the system. EOF will be in the same direction. This comigration condition often makes for separations in as little as 2 minutes. The feasibility of separating metal cations is readily predictable from a table of limiting equivalent conductances. Table 10.3 predicts an easy separation of Li^+, Na^+ and K^+ but the similarity in the conductances in NH_4^+ and K^+ makes their separation difficult to impossible. Magnesium(II) and the alkaline earths can be separated as the free ions except for Ca^{2+} and Sr^{2+}. The divalent metal ions listed cannot generally be separated as the free ions except for Pb^{++}, which has a higher conductance. The lanthanides all have very similar conductances and cannot be separated as the free ions.

Table 10.3. Limiting equivalent conductances λ (S cm^2 equiv^{-1}) of selected metal cations.

Ion	λ	Ion	λ	Ion	λ
Li$^+$	39	Mg^{2+}	53	Fe^{2+}	54
Na$^+$	50	Ca^{2+}	60	Co^{2+}	53
NH$_4^+$	73	Sr^{2+}	59	Ni^{2+}	54
K$^+$	74	Ba^{2+}	64	Cw^{2+}	55
Rb$^+$	78	Zn^{2+}	53		
Cs$^+$	77	Pb^{2+}	71		

Indirect detection is commonly used for metal cations because most of these cations lack the high UV or visible absorptivity needed for direct photometric detection. Waters Associates introduced UV Cat 1 for indirect detection. This is an amine cation that absorbs strongly in the UV spectral region and has an electrophoretic mobility similar to 1+ and 2+ metal cations. Protonated phenylethylamine or 4-methylbenzylamine are suitable visualization reagents for indirect detection of metal cations at moderately acidic pH values.

10.4.2 Separation of Free Metal Cations

The separation of K$^+$, NH$_4^+$, Na$^+$ and Li$^+$ using indirect photometric detection is demonstrated in Fig. 10.9a. At pH 6.15, K$^+$ and NH$_4^+$ are not separated, but at pH 8.5

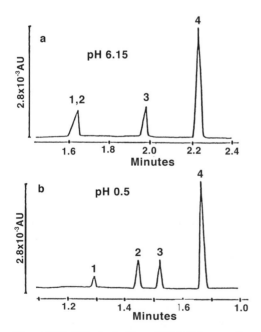

Figure 10.9. Effect of pH on NH$_4^+$–K$^+$ separation. 1 = potassium, 2 = ammonium, 3 = sodium, 4 = lithium (From Ref. [8], with permission).

a portion of the NH_4^+ has been converted to NH_3. The average charge is now <1 and the ammonium peak comes out later (Fig. 10.9b).

Incorporation of a crown ether (18-crown-6) and some methanol in the BGE improve the separation of metal ions [9].

In particular, 28-crown-6 complexes K^+, Sr^{2+}, Ba^{2+} and Pb^{2+}, causing them to migrate more slowly. Figure 10.10 shows a separation of Mg^{2+}, Ca^{2+}, Sr^{2+} and Ba^{2+} in the presence of a much larger amount of Na^+. Use of 18-crown-6 also makes it possible to separate NH_4^+, K^+ and several other common metal cations (Fig. 10.11).

Some metal cations absorb sufficiently in the UV spectral range to be detected directly. By working at a sufficiently acidic pH hydrolysis is inhibited and an additional buffer does not need to be added. Sharp peaks were obtained for UO_2^{2+} in an aqueous HCl electrolyte with direct detection at 214 nm [10]. Uranium(V) had not been separated previously as the uranyl ion.

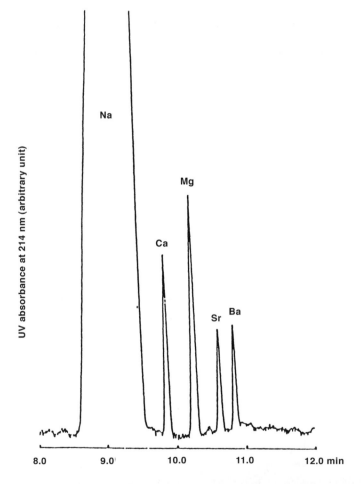

Figure 10.10. Effect of 18-crown-6 in the separation of 75 ppm Na^+ and 1 ppm Mg^{2+}, Ca^{2+}, Sr^{2+} and Ba^{2+}. Electrolyte, 15 mM lactic acid, 10 mM 4-methylbenzylamine, 32 % (v/v) methanol, 3.0 mM 18-crown-6, pH 4.3; applied voltage, 15 kV; injection time, 30 s. From Ref. [9], with permission.

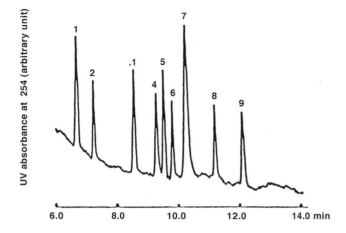

Figure 10.11. Electropherogram of a standard mixture with nine common cations using a nicotinamide electrolyte. Electrolyte, 8 mM nicotinamide, 12 % methanol, 0.95 mM 18-crown-6, pH 3.2 adjusted with formic acid; applied voltage, 25 kV; injection time, 40 s. Peaks: 1 = NH_4^+ (1.5 ppm); 2 = K^+ (1.5 ppm); 3 = Ca^{2+} (1.0 ppm); 4 = Na^+ (1.0 ppm); 5 = Mg^{2+} (0.5 ppm); 6 = Sr^{2+} (1.0 ppm); 7 = Al^{3+} (1.0 ppm); 8 = Ba^{2+} (1.0 ppm); 9 = Li^+ (0.2 ppm). From Ref. [9], with permission.

Iron(III) initially gave a rather broad peak with some tailing at t_m = 11.8 min when run in HCl at pH 2.4. However, a much sharper peak for iron(III) was obtained by increasing the chloride content of the capillary electrolyte to actually 66 mM and maintaining a 3- to 6-fold ratio of chloride in the electrolyte to that in the sample. An excellent separation of UO_2^{2+} and Fe^{3+} was obtained. Several other metal ions were separated at low pH with direct detection including vanadium(IV) and (V), VO^{2+} and VO_2^+ [10].

10.4.3 Separations Using Partial Complexation

Our ability to separate free metal cations by CE is limited because many of the metal ions have similar electrophoretic mobilities. An excellent way to enhance the separation of metal ions is to add a relatively weak complexing ligand (L^-) such as tartrate, lactate or α-hydroxyisobutyric acid (HIBA) to the BGE. Now part of each metal ion will remain as the free ion (M^{2+}, for example) and part will be converted to a complexed form (ML^-, ML_2, ML_3^-, for example). The total mobility (μ) will be the sum of the mole fraction of each species (α) multiplied by its mobility.

$$\mu = \alpha_M\mu_M + \alpha_{ML}\mu_{ML} + \alpha_{ML2}\mu_{ML2} + ... + \mu_{os} \qquad (10.3)$$

μ_{os} is the electroosmotic mobility. The free metal ion will make the greatest contribution to total mobility because of its higher positive charge. Different elements will in general be complexed to different degrees so that their net mobilities will vary even though the mobilities of uncomplexed cations may be almost the same.

Jones et al. obtained excellent separation of 15 alkali, alkaline earth, and divalent transition metal ions with 6.5 mM HIBA at pH 4.4 to partially complex some of the cations [11]. A protonated amine cation containing a benzene ring (Waters UV Cat 1) was used for indirect UV detection. All of the 13 lanthanides have been separated using HIBA under similar conditions (Fig. 10.12).

Lactate has the same α-hydroxycarboxylate complexing group as tartrate and HIBA, but it is a smaller molecule and forms somewhat weaker complexes than tartrate with most metal ions. Shi and Fritz found that a lactate system gave excellent separations for divalent metal ions and for trivalent lanthanides. A brief optimization was first carried out to establish the best concentrations of lactate and UV probe ion and the best pH. Excellent separations were obtained for all thirteen lanthanides, alkali metal ions, magnesium and the alkaline earths, and several divalent transition metal ions. All of these except copper(II) eluted before the lanthanides. An excellent separation of 27 metal ions was obtained in a single run that required only 6 min (Fig. 10.13).

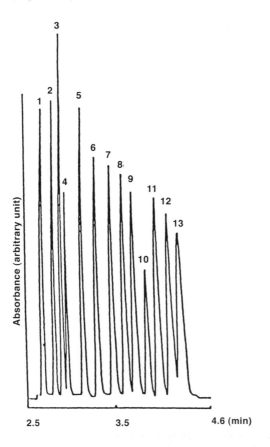

Figure 10.12. Separation of 13 lanthanides using HIBA. Electrolyte: 4 mM HIBA, 5 mM UV Cat 1, pH 4.3. Applied voltage: 30 kV. Peaks: 1 = La^{3+}; 2 = Ce^{3+}; 3 = Pr^{3+}; 4 = Nd^{3+}; 5 = Sm^{3+}; 6 = Gd^{3+}; 7 = Tb^{3+}; 8 = Dy^{3+}; 9 = Ho^{3+}; 10 = Er^{3+}; 11 = Tm^{3+}; 12 = Yb^{3+}; 13 = Lu^{3+} (From Ref [9] with permission).

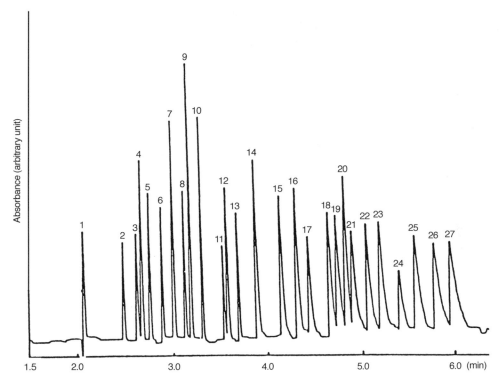

Figure 10.13. Separation of 27 alkali, alkaline earth, transition, and rare earth metal ions in a single run with lactate. Electrolyte: 15 mM lactic acid, 8 mM 4-methylbenzylamine, 5 % methanol, pH 4.25. Applied voltage: 30 kV. Peaks: 1 = K^+; 2 = Ba^{3+}; 3 = Sr^{3+}; 4 = Na^+; 5 = Ca^{2+}; 6 = Mg^{2+}; 7 = Mn^{2+}; 8 = Cd^{2+}; 9 = Li^+; 10 = Co^{2+}; 11 = Pb^{2+}; 12 = Ni^{2+}; 13 = Zn^{2+}; 14 = La^{3+}; 15 = Ho^{3+}; 16 = Pr^{3+}; 17 = Nd^{3+}; 18 = Sm^{3+}; 19 = Gd^{3+}; 20 = Cu^{2+}; 21 = Tb^{3+}; 22 = Dy^{3+}; 23 = Ho^{3+}; 24 = Er^{3+}; 25 = Tm^{3+}; 26 = Yb^{3+}; 27 = Lu^{3+} (From Ref [9] with permission).

It was not possible to separate ammonium and potassium by lactate alone. However, a mixture of lactate and 18-crown-6 permitted the separation of NH_4^+ and K^+ as well as most of the other ions in Fig. 10.13.

10.4.4 The Separation Mechanism

With partial complexation, each species of a given metal will migrate at a rate proportional to its mobility. The free metal ion will have a greater mobility than the various complexed species. This being the case, how are we able to obtain a sharp peak for each metal, as indicated in the separations just described? The answer is that a rapid equilibrium occurs so that each metal will move through the capillary as a tight zone. For example, M^{2+} will initially migrate faster than a complexed species ML^+, but as soon as M^{2+} moves ahead it encounters more free ligand and rapidly equilibrates to form more ML^+. At the back edge of the zone, the slower moving complexes reequilibrate to give a greater fraction of M^{2+}.

Further insight into the separation is given by the separation of lanthanides with 4 mM HIBA at pH 4.3 as the complexing agent (Fig. 10.12). Using published formation constants, the fraction of rare earths present in various chemical forms was calculated by a well-known method under the same conditions of pH and HIBA concentration used for the CE separation in Fig. 10.12. The calculated distribution of chemical species for each rare earth is shown in Table 10.4.

Table 10.4. Fractions of free (α_M) and complexed (α_{ML3}) rare earth metal ions and average number of ligands (n) in 4 mM HIBA electrolyte solution at pH 4.3.

Metal	α_M	α_{ML}	α_{ML2}	α_{ML3}	α_{ML4}	n
La	0.578	0.360	0.612			0.482
Ce	0.448	0.496	0.052	0.004		0.612
Pr	0.407	0.496	0.093	0.005	0.000	0.697
Nd	0.333	0.572	0.085	0.010	0.000	0.772
Sm	0.296	0.520	0.170	0.013	0.001	0.903
Gd	0.250	0.481	0.244	0.024	0.001	1.045
Tb	0.181	0.470	0.307	0.040	0.002	1.212
Dy	0.141	0.384	0.387	0.084	0.004	1.426
Ho	0.122	0.365	0.413	0.093	0.006	1.494
Er	0.097	0.309	0.472	0.112	0.010	1.629
Tm	0.079	0.309	0.473	0.123	0.016	1.686
Yb	0.070	0.296	0.431	0.169	0.033	1.797
Lu	0.047	0.222	0.514	0.172	0.045	1.946

The predominating species are the free metal ion (M^{3+}), the 1:1 complex (probably ML_2^+), the 2:1 complex (probably ML^{2+}), and the 3:1 complex (probably ML_3). A small fraction of the higher rare earths is also present as the 4:1 complex. Another striking feature is that the average number of ligands associated with a rare earth (n) increases rapidly with increasing atomic number. This occurs in a linear manner as demonstrated by a plot of n against atomic number.

The proposed mechanism necessitates a very fast rate of equilibrium between the free metal ions and the various complexed species. This condition is fulfilled with lactate and HIBA for the metal ions studied. However, metal ions that have slow complexation kinetics cannot be determined by a partial complexation CE system. For example, aluminum(III) gave no peak in a lactate system.

10.4.5 Separation of Organic Cations

In acidic solution organic amines can be separated by CE as the protonated cations. If necessary, some methanol or acetonitrile may be mixed with the aqueous BGE to enhance the solubility. For example, amino acids are zwitterions throughout much of the pH range but they have a net positive charge at very low pH values. The CE separation of a mixture of twenty common amino acids at pH 2.8 with direct detection at 185 nm has been described [13]. Isomeric anilines have been separated at pH 3.45 or 3.7 in a BGE containing 7.5 % 2-propanol [14].

Yeung and Lucy were able to obtain isotopic separations of ^{15}N-labeled amines from the natural ^{14}N-amines [15]. Their approach was to increase $\Delta\mu$ relative to the overall mobility:

$$R = \frac{\sqrt{N}}{4} \frac{\Delta\mu}{\mu + \mu_{eo}} \tag{10.4}$$

where N is the efficiency, μ is the mean mobility of the solutes and μ_{eo} is the electroosmotic mobility. The above equation states that when the solute mobility is counter balanced by the opposing electroosmotic flow (EOF), resolution approaches maximum and ultra-high resolution separations result. Their approach was to precisely control the EOF by use of mixed zwitterionic and cationic surfactants as buffer additives. The separations were based on the small isotopic effect on the dissociation constant.

10.5 Combined Ion Chromatography–Capillary Electrophoresis

10.5.1 Introduction

Perhaps the ultimate method for separation of ions would be one that combines the principles of ion chromatography and capillary electrophoresis into a single technique. CE has unusually high efficiency for separations of ions with at least some difference in ionic mobility. IC is based on differences in the sample ions for ion exchange sites in competition with the eluent ion. The migration order of ions through a capillary or ion exchange column may be different. Thus in CE the migration order of the halides is: $I^-, Br^- > Cl^- > F^-$. In IC the migration order is reversed:

$F^- > Cl^- > Br^- > I^-$.

An easy way to separate ions on the basis of both their electrophoretic and ion-exchange behavior is simply to add a water-soluble ion-exchange polymer to the BGE in a conventional CE setup. This idea was originally proposed by Terabe and Isemura [16,17] and was extended somewhat by Cassidy and his students [18,19]. A comprehensive paper by Li, Ding and Fritz [20] used poly(diallyldimethylammonium chloride) (PDDAC) as the soluble anion exchanger (see Fig. 10.14) in a BGE containing a

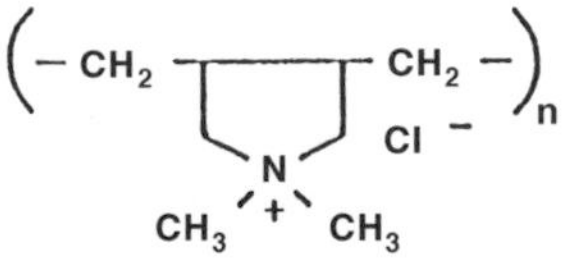

Figure 10.14. Structure of PDDAC. Molecular weight: medium: 200 000–350 000; high: 400 000–500 000.

relatively high concentration of a salt such as sodium chloride or lithium sulfate for the separation of inorganic and organic anions. The name "IC-CE" was coined for separations of this type.

10.5.2 Theory

The ion-exchanger equilibrium between a sample anion (A–) and the polymer ion exchange (P$^+$Cl$^-$) is given by the following equation:

$$P^+Cl^- + A^- \rightleftarrows P^+A^- + Cl^- \tag{10.5}$$

for which the equilibrium constant (K) is:

$$K = \frac{[P^+A^-][Cl^-]}{[A^-][P^+Cl^-]} \tag{10.6}$$

At a fixed concentration of P$^+$Cl$^-$, a conditional constant, K', may be written as follows:

$$K' = K\,[P^+Cl^-] \tag{10.7}$$

Combining Eqs. 10.6 and 10.7, and rearranging gives:

$$K = \frac{[A^-]}{[P^+A^-]} = \frac{[Cl^-]}{K'} \tag{10.8}$$

The migration rate of a sample anion will be proportional to the ratio of [A$^-$]:[P$^+$A$^-$]. The fraction of sample anion present as the free anion (A$^-$) will migrate rapidly toward the anode, while the fraction associated with the ion exchanger (P$^+$A$^-$) will move but slowly in the opposite direction. These equations show that salt concentration in the BGE (Cl$^-$ in the example) will have a major effect on sample analyte migration as well as the polymer ion concentration and the equilibrium constant, K.

Addition of 0.05 % to 0.30 % PDDAC to the BGE sets up a dynamic equilibrium in which PDDAC forms a thin coating on the inner walls of the capillary. This imparts a negative charge to the surface and sets up an electroosmotic flow toward the anode which is in the opposite direction to the usual cathodic EOF in uncoated capillaries. Under typical conditions the EOF in capillaries equilibrated with PDDAC was almost constant over a wide pH range (Table 10.5).

A negative power supply (–10 kV) is used for anion separations. The BGE typically contains 0.05 % to 0.30 % PDDAC, up to 150 mM sodium chloride or lithium sulfate, and a 20 mM borate buffer. The EOF vector and the electrophoretic vectors of the sample anions are both in the anodic direction. The fraction of each sample anion that is attached to exchange sites in the PDDAC has a cathodic electrophoretic vector, although this is believed to be weak. The net electrophoretic vector for any given anion depends primarily on the fraction that exists as the free ion (Eq. 10.7) and on the electrophoretic mobility of the free ion.

Separations of common inorganic anions are quite fast, as indicated by the migration times in Table 10.5. It is also possible to obtain a baseline because of the greater ion-exchange affinity for iodide; the electrophoretic mobilities of bromide and iodide are almost identical. A separation of several inorganic anions is illustrated in Fig. 10.15.

Table 10.5. Comparison of EOF and migration times at different pH values for inorganic anions. Capillary: 40 cm * 50 μm; electrolyte: 150 mM Li_2SO_4, 0.05 % PADDC, 20 mM borate for pH 8.5, or 20 mM sodium acetate for pH 5.0, or 20 mM HCl for pH 2.3; separation voltage: –10 kV; hydrostatic sampling: 40 s at 10 cm height; detection: UV: 214 nm; EOF marker: D.I. H_2O.

		Migration time (min)			
pH	EOF ($cm^2/V \cdot s$)	Br^-	I^-	NO_3^-	SCN^-
2.3	-2.46×10^{-4}	1.98	2.06	2.17	2.34
5.0	-2.65×10^{-4}	1.95	2.03	2.13	2.30
8.5	-2.74×10^{-4}	1.93	2.00	2.10	2.26

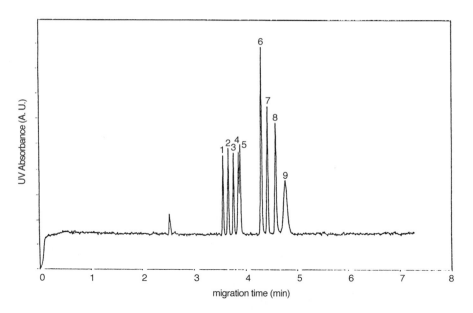

Figure 10.15. Separation of organic anions. Capillary: 40 cm * 50 μm; electrolyte: 150 mM Li_2SO_4, 20 mM borate, 0.3 % PADDC, pH 8.5; separation voltage: –10 kV; hydrostatic sampling: 40 s at 10 cm height; detection: UV, 214 nm. Peaks: 1 = benzenesulfonic acid; 2 = benzoic acid; 3 = *p*-toluenesulfonic acid; 4 = *p*-hydroxybenzoic acid; 5 = *p*-aminobenzoic acid; 6 = 2-naphthalenesulfonic acid; 7 = 1-naphthalenesulfonic acid; 8 = 3,5-dihydroxybenzoic acid; 9 = 2,4-dihydroxybenzoic acid.

10.5.3 Effect of Variables

Equation 10.7 predicts that increasing concentrations of PDDAC will result in slower migration (longer migration times) by reducing the fraction of an analyte that is present as the free anion. This is born out in practice. Concentrations of PDDAC as high as 0.8 % could be used.

As predicted by Eq. 10.7, a higher salt concentration in the BGE will decrease the ion-exchange effect and lead to shorter migration times. The type of salt as well as its concentration can have a major effect on a separation. Sulfates generally repressed the ion exchange effect more than chlorides. Comparison of separations at 50, 100, and 150 mM lithium sulfate showed much sharper peaks at the higher salt concentrations. This may be due to more efficient electrostacking. At 150 mM lithium sulfate, the average plate number for bromide, thiocyanate, chromate and molybdate was 122 000.

10.5.4 Scope

Organic anions as well as inorganic anions can be separated by IC-CE. Organic anions are generally more bulky than inorganic anions and therefore migrate more slowly. Figure 10.16 shows a separation of 17 inorganic and organic anions in a single run. It is unlikely that an equivalent separation of such a broad range of sample ions

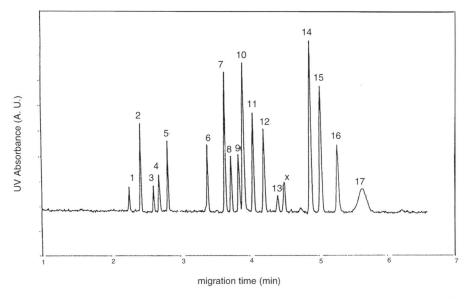

Figure 10.16. Separation of 17 inorganic and organic anions. Conditions same as Fig. 10.15. Peaks: 1 = bromide; 2 = nitrate; 3 = chromate; 4 = iodide; 5 = molybdate; 6 = phthalate; 7 = 1,2,3-tricarboxylate; 8 = 1,2-benzenedisulfonate; 9 = terephthalate; 10 = isophthalate; 11 = benzoate; 12 = *p*-toluenesulfonate; 13 = 1,2,5-tricarboxylate; 14 = 2-nathphalenesulfonate; 15 = 1-naphthalenesulfonate; 16 = 3,5-dihydroxybenzoate; 17 = 2,4-dihydroxybenzoate; x = unidentified impurity.

could be obtained by IC or CE alone. The ion-exchange effect is again apparent. The 1- and 2- isomers of naphthalenesulfonic acid are well resolved. In the absence of PDDAC these ions coelute.

In should be noted that direct photometric detection has been used for the separations described here. Conditions for indirect detection of samples as complex as those in Fig. 10.16 have not yet been worked out.

Separations of cations by IC-CE should also be feasible. This would require the use of a water-soluble polymer with sulfonate or carboxylate groups. The polymer should be transparent in the UV spectral region so that photometric detection could be used for the sample cations.

The fact that soluble polymers may undergo ion-exchange in solution may have an impact on future developments in ion separations. Ion exchange reactions in a single phase should avoid the inevitable peak broadening that occurs when the ion exchanger is a separate, solid phase.

References

[1]   F. E. P. Mikkers, F. M. Everaerts and T. P. M. Verheggen, Concentration distributions in free zone electrophoresis, *J. Chromatogr.*, 169, 1, 1979.
[2]   S. A. Shamsi and N. D. Danielson, Naphthalenesulfonates as electrolytes for capillary electrophoresis of inorganic anions, organic acids, and surfactants with indirect photometric detection, *Anal. Chem.*, 66, 3757, 1994.
[3]   K. M. Lau, PhD Thesis, University of Hong Kong, December 1997.
[4]   C. A. Lucy and T. L. McDonald, Separation of chloride isotopes by capillary electrophoresis based on the isotope effect on ion mobility, *Anal. Chem.*, 67, 1074, 1995.
[5]   M. J. Thornton and J. S. Fritz, Separation of inorganic anions in acidic solution by capillary electrophoresis, *J. Chromatogr. A.*, 770, 301, 1997.
[6]   W. Ding, M. J. Thornton and J. S. Fritz, Capillary electrophoresis of anions at high salt concentrations, *Electrophoresis*, 19, 2133, 1998.
[7]   G. Okafo, G. Harland and P. Camilleri, Enhanced resolution in the fingerprinting of short-chain oligonucleotides using the dodecasodium salt of phytic acid in capillary electrophoresis, *Anal. Chem.*, 69, 1982, 1996.
[8]   A. Weston, P. Brown, P. Jandik, W. R. Jones and A. L. Heckenberg, Factors affecting the separation of inorganic metal cations by capillary electrophoresis, *J. Chromatogr.*, 593, 289, 1992.
[9]   Y. Shi and J. S. Fritz, New electrolyte systems for the determination of metal cations by capillary zone electrophoresis, *J. Chromatogr. A*, 671, 429, 1994.
[10] M. J. Thornton and J. S. Fritz, Separation of metal cations in acidic solution by capillary electrolysis with direct and indirect UV detection, *J. High Resol. Chromatogr.*, 20, 653, 1997.
[11] W. R. Jones, P. Jandik and R. Pfeifer, Capillary ion analysis, an innovative technology, *Am. Lab.*, 5, 40, 1991.
[12] Y. Shi and J. S. Fritz, Separation of metal ions by capillary electrophoresis with a complexing electrolyte, *J. Chromatogr.*, 640, 473, 1993.
[13] R.L. Williams and G. Vigh, Maximization of separation efficiency in capillary-electrophoretic chiral separations by means of mobility matching background electrolytes, *J. Chromatogr. A.*, 730, 273, 1996.

11 Chemical Speciation

11.1 Introduction

The term chemical speciation refers to the identification and determination of the various chemical forms and individual amounts of an element that are present in a particular sample. Several elements such as arsenic and selenium are converted to different species during weathering, uptake or elimination through biological processes, fixation in sediments, or remobilization. Drinking water may contain different oxyhalide species formed through various water disinfection processes. Several elemental species of chlorine and bromine in drinking water are now regulated in the United States and Europe. Industrial samples such as plating baths contain different oxidation states of metals. The effectiveness of chromium plating baths depends on the amounts of Cr(III) and Cr(VI) that are present. Gold plating processes involve the use of Au(I) and Au(III) cyanide complexes. Vanadium speciation is important in the recovery of sulfur from geothermal water.

Environmental applications are perhaps of greatest interest. Speciation is important because each inorganic species of a particular element may possess quite different biological, medicinal or toxicological properties. There are differences of element-specific species in reactivity, biological availability and element transport in the environment and into the food chain. Knowing the chemical forms of an element in environmental, agricultural, or other samples can be much more important than knowing the total elemental content. For example, pollution incidents involving mercury have shown that total metal data are insufficient and often misleading in assessing the potential hazard of this metal. The general application of ion chromatography (IC) to environmental samples has been described in a review [1].

In order for a speciation analysis to be successful, the analytes must be stable throughout the entire analysis process. Most difficult can be keeping the sample stable up to the point it is injected into the instrument. Since the analytes can be oxidized or sometimes reduced to related species within the sample, simply handling the sample may result in changes in the relative concentrations of the analytes. Preconcentration of the samples is not advised and in many cases, the samples must be treated to preserve the original state. For example, EDTA may be added in some cases to complex the metals in a sample, preventing them from oxidation or reduction reactions. Nitric acid, which is normally used to preserve metal-containing samples, may not be used for speciation because nitric acid is an oxidizing agent. The following concerns should be addressed in the method-development stages. Is the sample photoreducible or photosensitive? Are any analyte species volatile or thermally unstable? Are there changes in redox conditions due to sampling, sample preservation, or sample-preparation techniques that are likely to affect the balance of analyte species? The loss or conversion of non-analyte sample components may also affect the balance of analyte species. The reader is advised to consider the chemistry closely and follow established speciation methods to prevent incorrect measurements of the various elemental species.

11.2 Detection

There have been several reports where general detection methods such as conductivity or post-column reaction photometry have been used for speciation detection. EPA method 300 describes the use of conductivity detection for the speciation of chloride and bromide in drinking water. The chromatographic conditions are much more important when general detection is used. Major interferences must elute well away from the ions of interest and all of the species should elute in the same chromatogram. The ionic form of the metal species must be either all positive or negative in order to use either cation chromatography or anion chromatography. This is not necessarily easy to arrange for some metals. Chromium can exist as either a cation or anion depending on the oxidation state.

The use of conductivity to speciate samples is described in many other sections in this book. This section describes the use of selective detection. Detectors that are selective or can be "tuned" to an element are quite useful. For example, it was shown in a review by Urasa [2] on DC plasma atomic emission spectrometry (DCPAE) detection that only one of the metal species need be retained by the chromatographic column in order to perform a speciation analysis. Cr(III) exists as a cation in aqueous solution. Cr(VI) exists as an anion, i.e., CrO_4^{2-}. If one is using a cation exchange column and injects a mixture of Cr(III) and Cr(VI), then the anion, Cr(VI), elutes with the void volume and Cr(III) is retained and elutes later. Selective detection allows quantification of the Cr(IV) peak in spite of the fact that many other materials are eluting at the same time. Cr(III) is eluted later as part of the cation-exchange process and detected. The eluent conditions needed to elute this peak are not very stringent. All that is needed is to elute Cr(III) quickly in a nice sharp peak after the Cr(VI). Since the eluent driving and counterions are not detected, the eluent type, pH, and/or concentration can be changed as needed. Of course, the reverse could be used if an anion-exchange column is used. Cr(III) would elute quickly and then Cr(VI) will follow. Thus, methods for speciation can be developed quickly and be performed easily provided there is access to selective detection.

Strong chelating or complexing agents may form anionic complexes with metals. These anionic complexes can be separated by anion-exchange chromatography. Continuing with the chromium example, Cr(III) can be converted to an anionic complex by complexation with EDTA. Thus, a mixture of CrEDTA and CrO_4^{2-} may be separated by anion-exchange chromatography and be detected by UV detection.

In this section, many types of selective detection are referenced. However, inductively coupled plasma mass spectrometry (ICP-MS) appears to be the most useful, because it is very selective and sensitive. In many cases, the methods described for one form of detection can be translated into other selective detection methods, provided the sensitivity is high and interference from other analytes is low. Inductively coupled plasma coupled to atomic emission (ICP-AE) is also a powerful and sensitive detection method for elemental speciation.

11.3 Chromatography

The majority of speciation separations are performed by ion-exchange chromatography although there has been significant work performed by ion-pair chromatography. The ion-exchange separation of metal species in a chromatographic system can be performed on the basis of either of two concepts. Sample species can be separated on a column on the basis of affinity differences between the species for the column. Sample species may also be separated by a complexing reagent in the eluent. The complexing reagent changes the form of the sample species; this allows it to move down the column more easily.

These concepts are sometimes called the "push/pull" mechanism of ion-exchange chromatography [3]. The eluent that competes with sample species for ion-exchange sites and elutes the sample from the column operates under a "push" mechanism. The eluent "pushes" or elutes the samples species from the column. Adding a complexing reagent to the eluent and converting the sample species into a complex that is not as well retained by the ion-exchange column is called the "pull" mechanism. The complexing agent "pulls" the sample species down the column.

The effect can be shown by equations. It has been shown that the ion-exchange reaction of the eluent and sample species competing for the ion exchangers can be written as an equilibrium equation (Chapter 5). By chromatographic theory, the equilibrium equation can be rearranged to predict behavior in ion-exchange chromatography. The result is shown by the equation:

$$\log k = a/e \log C - a/e \log E + 1/e \log K_{eq} - a/e \log e + D$$

where k is the sample capacity factor, a is the charge of the sample ion, e is the charge of the eluent-driving ion, C is the resin capacity, E is the eluent concentration, and K_{eq} is the static solution ion-exchange equilibrium constant. The constant D includes the terms column void volume and resin density.

Examination of the equation shows that controlling separations (changing $\log k$ or retention) by affinity differences is not very powerful. Options for controlling the separation are limited. The resin capacity, sample ion charge, and affinity of a sample (K_{eq}) for a particular column is usually constant or not easily changed. The only real variables are the eluent type (affinity of eluant ion for the ion exchanger), eluent ionic charge, and the eluent pH and concentration. Some retention crossovers can be achieved by the use of different eluent types and variations in the eluent concentration.

But, the addition of a complexing agent to the eluent makes ion-chromatographic separations really powerful. The complexing agent controls the amount of ion species available to compete with the eluent for the ion exchanger. In effect, the K_{eq} term is changed (usually lowered). In cation chromatography, K_{eq} (and log k) is usually lowered because neutral or anionic complexes are formed. In anion chromatography, K_{eq} and retention is usually increased. The complexing agent is usually specific for a metal or group of metals. Thus, a particular separation can be achieved by choosing the type, concentration and pH of the complexing agent added to the eluent.

11.4 Valveless Injection IC

Normal IC separations rely on a high-pressure eluent pump, injection valve, column and detector. However, a novel method for speciation was developed by Gjerde and Wiederin [4], in which a low-pressure column, valveless injection method was employed. The system configuration is shown in Fig. 11.1. An example with this type of IC to perform a separation of Cr(III) and Cr(IV) is shown in Fig. 11.2. Nitric acid eluent is acceptable in this case because it does not oxidize Cr(III). The separation method is anion exchange, so Cr(III) is unretained and Cr(IV) is eluted by the nitrate

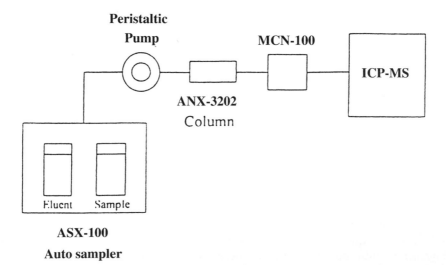

Separation Schematic

Figure 11.1. Schematic representation of a valveless, low pressure column, IC. The ASX-100 Autosampler, ANX3202 anion exchange column, and MCN-100 low flow volume nebulizer are products of Transgenomic, Omaha, NE (Courtesy of Transgenomic, Inc.).

1 ppb Cr (III) and Cr (VI)

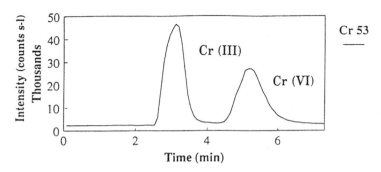

Figure 11.2. Separation of Cr(III) and Cr(VI) by valveless IC and ICP-MS detection. The separation is affected by both the volume of sample introduced to the column and the speed of the peristaltic pump. The eluent was 0.35 % (w/w) nitric acid adjusted to pH 1.6 with ammonium hydroxide. The column is a low-capacity anion exchanger ANX3202 with dimensions of 3.2 × 20 mm. Detection limits for Cr(III) and Cr(VI) are both <0.1 ppb. Courtesy of Transgenomic, Inc., Omaha, NE.

driving anion. Fig. 11.3 shows a separation of As and Se species with a valveless injection IC. Since the separation is performed under the same elution conditions and the detection by inductively coupled plasma mass spectrometry (ICP-MS) is multichannel, detection is simultaneous for the two elemental species.

The sample probe is moved between eluent and standard or sample solutions using an autosampler. Some wells in the autosampler contain samples to be injected. Other

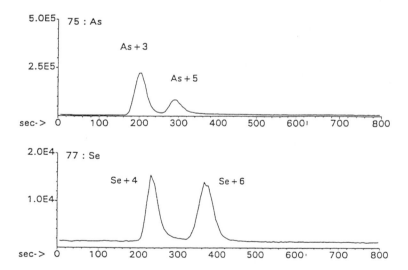

Figure 11.3. Speciation of one 50 μL injection of 100 μg/L of each of the ions As(III), As(V), Se(IV), and Se(VI) by Cetac's ANX3202-As/Se chromatography column with 5 mM ammonium malonate eluent (pH 8.5). The column is a low-capacity anion exchanger with dimensions of 3.2 × 20 mm. Nebulization: CETAC MCN-100 Micro Concentric Nebulizer. Detection: ICP-MS, with 75 mass isotope for As and 77 mass isotope for Se. Courtesy of Transgenomic, Inc., Omaha, NE.

wells contain standards of known concentration. Still others contain one or more eluents of different concentration.

A peristaltic pump delivers the eluent and sample to the low-pressure column. Because a peristaltic pump is a flow-through device, the integrity of the sample is maintained as it is passed through the pump. However, because peristaltic pumps can only pump at a maximum of 100 psi, the columns must operate at a low backpressure. Development of a low backpressure column depends on several factors. First, the ion exchange resin must be small, uniform and efficient. The column hardware and fluid connections must be well designed and not contain flow paths that broaden the sample peaks. The column should be short so that the eluent backpressure is low. Finally, the detector must be able to operate at low (eluent) flow rates. The low eluent flow rates (approximately less than 100 μL/min) generate lower backpressures than standard flow rates.

The column is designed to provide a moderate plate count (approximately 500 plates) at a low flow rate (typically 75 μL/min) and a low backpressure (typically 50 psi). After the column, the flow is directed to a Cetac MCN-100 nebulizer, designed to operate at low flow conditions, and then to an inductively coupled plasma and finally to a mass spectrometer.

The injection process is as follows: during the process of injection, the probe is moved from the eluent well into the sample well for a fixed period of time and then back to the eluent well. Because a finite amount of time is needed to travel from the eluent to the sample and back again, an air segment is introduced on each end of the sample. This does not harm the separation, as long as the time interval is not very long. In fact, an air segment serves to keep the sample integrity so that it is injected cleanly into the column. A typical time interval is less than 1 s, so for a flow rate of 75 μL/min, the air segment volume is just over 1 μL. Sampling time is about 30 s. A longer sampling time gives higher sensitivity, but worse chromatographic resolution. A shorter sampling time may be used for high sample concentrations. The same sampling time should be used for both samples and standards.

One of the advantages of the valveless mode is the ability to perform step-gradient elution. An example of this is shown in Fig. 11.4. In this case, the eluent is pH 9.0 malonic acid, used at concentrations of 1 to 50 mM. At 600 s, the concentration is stepped from 1 mM to 5 mM and at 1000 s the concentration is stepped to 50 mM. The step gradients are formed by simply changing to a different well in the autosampler. This method, used on a NIST sample, is shown in Table 11.1.

Since ICP-MS detection is used, a non-analyte element may be spiked into the eluent to provide a known, uniform signal for instrument optimization. Alternatively, the column may be removed from the sample-introduction apparatus to allow direct pumping of optimization solutions to the detector. Another advantage of removing the column is that the total element content of a particular species can be determined and compared to the total amount of the various species. This could be done automatically with a six-port, two-position switching valve where the column is positioned in one of the bypass loops. Each sample could be analyzed with and without the column inline. This, in fact, brings up an interesting point. Essentially, the only difference between this type of chromatograph and a normal ICP-MS instrument is the presence

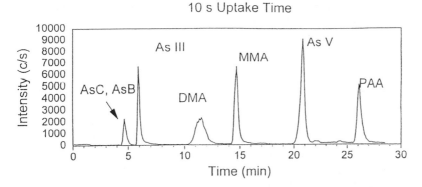

Figure 11.4. Chromatogram of step-gradient, valveless IC. 10 µg/L spike of NIST 2670, elevated level diluted 50× with 5 µM malonic acid, pH = 9, adjusted with ammonium hydroxide with flow rate 75 mL/min. Arsenic compounds: AsC (arsenocholine), AsB (arsenobetaine), As(III) (arsenite), DMA (dimethylarsenic), MMA (monomethylarsenic), As(V) (arsenate), PAA (phenylarsonic acid). HP4500, Cetac M1 MCN. The column is a low-capacity anion exchanger ANX3206 with dimensions of 3.2 × 60 mm. Courtesy of Transgenomic, Inc., Omaha, NE.

of a column and the types of solvents (eluents) used to introduce the sample. This point may help the valveless IC gain acceptance by the analytical spectroscopist.

Table 11.1. Arsenic Species found in NIST SRM 2670.

Species	Elevated level (µg/L)
AsB/AsC	13.6
As(III)	8.3
DMA	44.2
MMA	11.6
As(V)	374
Total Found	452
Certified Value	480 ± 100

11.5 Speciation of Metals

11.5.1 Chromium

Chromium is known in all oxidation states between 0 and VI, but is commonly found in oxidation states III and VI. Cr(III) species in acidic solution exist as $Cr(H_2O)_6^{3+}$ ions and in concentrated alkali have been identified as $Cr(OH)_6^{3-}$ and $Cr(OH)_5(H_2O)^{2-}$. These ions are regular octahedra. Cr(VI), in basic solution above pH 6, exists as tetrahedral yellow chromate, CrO_4^{2-}, ion. Between pH 2 and 6, $HCrO_4^-$ and the orange-red dichromate, $Cr_2O_7^{2-}$, are in equilibrium. Under strongly acidic conditions, only dichromate ion exists. Addition of alkali to dichromate results in chromate.

Chromium metal is produced by roasting the chromate ore. The largest single use of the metal is as corrosion inhibitors in alloying steels. Chromium finds its way into the environment through industrial wastes from electroplating sludge, tannery wastes, the manufacture of corrosion inhibitors, and municipal sewage sludge. These sludges are of particular concern. Cr(III) is essential to human nutrition. Cr(III) is present in most soils while Cr(VI) is only occasionally present. Cr(III) is less toxic and less mobile (in the environment) than Cr(VI). Cr(VI) compounds are very toxic to aquatic plants and animal life as evidenced by their widespread use as algicides. Cr(VI) toxicity may manifest itself in the form of skin ulceration, nasal perforation, and lung cancer.

There have been several reports [6–10] that describe the speciation of chromium. Many of these reports described the use of DC plasma atomic emission (DCPAE), ICP-AE, and atomic absorption (AA) selective-type detection. UV/VIS detection has also been extensively used. Atomic emission detection has been found to be very promising due to its selectivity and low detection limits.

The basis of many of the separations has been to convert Cr(III) to an anion by adding a complexing agent. Cr(VI) is already an anion (usually CrO_4^-), hence, anion chromatography can be used to separate mixtures. An example is separations carried out in the presence of KSCN, converting Cr^{3+} to $Cr(SCN)_4^-$.

Pyridine dicarboxylic acid (PDCA) was used as a complexing agent for Cr^{3+} to form anionic $Cr(PDCA)_2^-$. The sample pH is critical to the separation. Optimum results were reported at pH 6.8. Both Cr(III) and Cr(VI) were detected by UV/VIS detection and diphenylcarbohydrazide as a post-column reagent. The method was found to be applicable for plating baths and waste-water analysis. Geddes and Tarter [9] reported the use of EDTA eluent and UV detection for Cr speciation. Two silica-based anion exchange columns were connected in series. However, the chromatographic peaks were very broad and the method had poor detection limits.

Much better chromatographic separations were shown by Powell et al. [11]. Spiking Cr(VI) into industrial waste-water samples showed no initial concentrations of Cr(III), but after 3 weeks, 30 % of the Cr(IV) had converted to Cr(III). Other work [8] shows that nitric acid is an effective preservative for determination of chromium in surface water.

At ambient pH, the unpreserved water exhibited over 50 % loss of Cr(III) within 1.5 hours, due to precipitation. With nitric acid preservation, both chromium species were preserved in solution for a week. Care must be taken when spiking standards into any natural water sample. Organisms and suspended solids present in the water are easily oxidized by Cr(VI). All possible sample matrices cannot be anticipated so acid (or other) preservation cannot be recommended as a general strategy. The best sample-handling protocol for specific needs and specific matrices must be determined during method development. Prompt analysis of the sample is often the best approach for accurate determinations.

11.5.2 Iron

Iron is the second-most abundant metal after aluminum. The highest oxidation state of iron is VI, although, of course, II and III oxidation states are the most common. Fe(II) forms a variety of complexes. In aqueous solution Fe(II) exists as $Fe(H_2O)_6^+$, which is pale sea-green in color. Fe(II) is slowly oxidized in acid. Most Fe(II) complexes are octahedral. Perhaps the most important ferrous iron complex to humans is heme which exists in hemoglobin.

In aqueous solution, Fe(III) has a strong tendency to hydrolyze. The hydrated ion complex $Fe(H_2O)_6^{3+}$ is pale purple and exists only in strongly acidic solutions (pH 0 or less). In less acidic media, hydroxy complexes are formed. In the presence of complexing anions such as Cl^-, the hydrolysis of Fe^{3+} or of $FeCl_3$ is more complicated, and chloro, aqua, and hydroxy species are formed. Fe(III) complexes are mainly octahedral just as they are for Fe(II). Fe(III) has a greater affinity for oxy ligands, whereas Fe(II) has a slight preference for ligands containing nitrogen-donor atoms.

The simultaneous determination of Fe(II) and Fe(III) is important to understanding the environmental redox processes in biological systems. Iron activity affects several chemical processes in natural waters and its speciation concentration is a significant factor in the evaluation of water quality.

Iron speciation of Fe(II) and Fe(III) is reported more often than any other speciation. The methods are based on cation-exchange, anion-exchange, and ion-pairing chromatography. Only a few of the methods are discussed here. Saitoh and Oikawa [12] simultaneously determined Fe(II) and Fe(III) by post-column reaction (PCR) detection. The PCR reagent was bathophenanthrolinedisulfonic acid and ascorbic acid. This procedure was found to be successful in spring-water samples. Moses et al. [13] reported the determination of Fe(II) and Fe(III) in water samples. The detection system consisted of PCR reaction with PAR detection reagent. In this work, the Fe(II)/Fe(III) ratio increased at times – probably due to Fe(III) being photochemically reduced.

The presence of trace iron contaminants in gold plating baths can cause brittle deposits. $Fe(CN)_4^-$ and $Fe(CN)_6^{3-}$ were reported to be separated by anion chromatography [14]. These complexes are multivalent and difficult to elute. Small amounts of Na_2CO_3 added to the mobile phase sharpened the peaks.

11.5.3 Arsenic

Arsenic is found in igneous and sedimentary rocks. The most common commercial source is as the by-product from the refining of copper, lead, cobalt, and gold ores. Although arsenic is actually a metalloid, it is grouped with metals for most environmental purposes.

Arsenic chemistry is complex, involving a variety of oxidation states, both as anionic and cationic species, and both inorganic and organometallic compounds. Of these, III and V are the most common oxidation states. The oxidation states of arsenic change easily and reversibly. As(III) is commonly encountered as the arsenite ion, $H_2AsO_3^-$. Arsenious acid is a weak acid, $pK_{a1} = 9.2$, $pK_{a2} = 13$.

As(V) exists as $H_2AsO_4^-$ in aqueous solution. Arsenic acid is a weak tribasic acid. Its dissociation constants, $pK_a = 2.3$, 6.8, and 11.5, are similar to those of phosphoric acid. Oxidation of As(III) to As(V) in dissolved oxygen is slow at neutral pH, but is much faster at either extreme [15]. In reducing environments, As(III) is produced, but As(V) is the most stable state in aerobic environments.

Most arsenic compounds are highly toxic, causing dermatitis, acute and chronic poisoning, and possibly cancer. Arsenic is found in virtually all soil and other environmental matrices [16]. Arsenic is present in coal, pesticides, preservatives, etc. Arsenite, a commercial form of arsenic, is one of the most toxic forms of arsenic.

Arsenic speciation by IC can be simple and reproducible. One procedure, conductivity detection of As(III) and As(V), was reported by McCrory-Joy [17]. The procedure was sensitive and there was no interference by ions such as NO_3^-, HPO_4^{2-}, and SO_4^{2-}, that are present in the sediment extract. However, detection of As(III) is not possible by suppressed conductivity because of its weak acid strength.

A method for several species of As is shown in Fig. 11.5. The method employs the Cetac AN1 column (Transgenomic, San Jose, CA) and ICP-MS detection (HP). The method has become standardized for the determination of As species in drinking water in Japan [18].

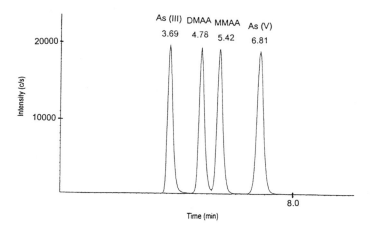

Figure 11.5. Chromatogram of 100 µg each of As(III), DMAA (dimethyl arsenic acid), MMAA (monomethyl arsenic acid), As(V) species for standard test of species in drinking water, by standard IC and ICP-MS detection. Column: AN1 (Cetac Technologies, Omaha, NE). Eluent: 2 mM PBS/0.2 mM EDTA, pH 6.0 at 1 mL/min. Chromatography courtesy of T. Matsuda, HP-Japan, 1999.

11.5.4 Tellurium

Tellurium is used in the metallurgical industry as an alloy constituent. Tellurium improves the acid resistance of lead used in batteries. It is also used in the manufacture of heat- and abrasive-resistant rubber. Tellurium is obtained as a by-product in the electrolytic refining of copper. In the semiconductor industry, the ultra-trace level determination of tellurium in tellurium-doped single crystals is often required. Tellurium species are highly toxic.

Tellurium and selenium resemble each other chemically. The analytical chemistry of both elements is usually presented together. Tellurium and selenium are commonly found in oxidation states II, IV, and VI, as well as in the elemental forms. Oxidation state IV is the most stable.

In aqueous chemistry, tellurium is mainly found as telluride (Te^{2-}), tellurite (TeO_3^{2-}), and tellurate (TeO_4^{2-}). Zolotov et al. [19] developed an IC procedure for the separation of TeO_3^{2-}, and TeO_4^{2-} by a suppressed ion-chromatographic system. In the method, F^- interfered with the determination of TeO_3^{2-}, while SO_4^{2-} interfered with the determination of TeO_4^{2-}. In a report by Chen et al. [20], a similar method was used to speciate tellurium. The detection limits for TeO_4^{2-} were very poor in both methods. The studies were confined only to standard solutions.

11.5.5 Selenium

Selenium is widely dispersed. It is found, for example, in igneous rocks, volcanic sulfur deposits, hydrothermal deposits, and copper ores. Selenium is used in the electronics industry for the manufacture of rectifiers and photoconductivity cells. Selenium and its compounds are also used as additives in chromium-plating, glass, ceramics, pigment, rubber, photography, lubricants, pharmaceuticals, and organic substances.

Selenium is both a toxic and an essential element. The toxicity depends greatly on the species. Selenium is a cumulative toxic substance and can be a serious health hazard when present in high concentrations in food and water. However, at very low levels, µg/kg it is recognized as an essential trace element in animal nutrition.

There has been an increasing interest in the determination of selenium at trace levels in a wide variety of matrices. Selenium determinations of environmental samples have become quite important. Large-scale poisoning of water fowl has occurred in several watershed areas of central California. Selenium is washed or leached into these areas, helped through the widespread irrigation of selenium-containing farm soil.

Selenium is commonly found in oxidation states II, IV, and VI, and as an element. Se(VI) is much more stable than Se(IV). The reaction chemistry of selenium is mainly that of selenide (Se^{2-}), selenite (SeO_3^{2-}), and selenate (SeO_4^{2-}). Selenious acid is a weak acid with a pK_a of 2.6 and 8.3. Selenious acid and selenite are much stronger oxidants than sulfurous acid and sulfite. Thus, many of the characteristic reactions are related to redox reactions, in which Se(IV) can be reduced to elemental selenium. The acid strength of selenic acid (H_2SeO_4) is similar to sulfuric acid.

Selenium can exist in at least two different ionic forms in environmental samples: selenite (SeO_3^{2-}) and selenate (SeO_4^{2-}). The concentration and speciation of selenium in a given sample depend on the pH and redox conditions, the solubility of its salts, the biological interactions, and the reaction kinetics. For example, in sea water, SeO_3^{2-} is the dominant species, but in river or tap water, selenium can be found in roughly equal Se(IV) and Se(VI) amounts [21].

Zolotov et al. [19] developed a suppressed IC method for the simultaneous determination of SeO_3^{2-}, and SeO_4^{2-} in the presence of F^-, Cl^-, NO_3^-, HPO_4^{2-} and SO_4^{2-}. The separation took about 30 minutes. The method was applied to river and tap water. Sensitivity and precision were good. Hydrogen peroxide was used in the sample pretreatment to decompose organoselenium compounds so that total selenium could be determined.

The determination of SeO_3^{2-} and SeO_4^{2-} in water with graphite furnace atomic absorption detection was investigated by Chakraborti et al. [22]. Some interference by Cl^- and F^- was reported. DCPAE detection was used by Urasa and Ferede [23]. Results were 1000 times more sensitive than conductivity detection. One of the advantages of atomic emission detection was described in this work. Identical molar sensitivity was obtained for both species. Mehra et al. [24] developed a novel single-column IC method to determine selenium species in seleniferous soil samples. The separation took about 14 minutes and there were no reported interferences.

11.5.6 Vanadium

Vanadium has a relative abundance of about 0.02 %. Oxidation states of V to I are known. Vanadium solutions generally contain several species in a complicated series of equilibria. V(V) and V(IV) are both stable, with the former mildly oxidizing and represented mainly by oxy species. Pervanadyl ion (VO_2^+) is a major species in strongly acidic solutions, while in strong base, the mononuclear vanadate, VO_4^{3-} exists. V(IV) ions are stable in acid, and give blue solutions of vanadyl ion (VO^{2+}). A number of anions of the IV oxidation state are known, including VO_3^{2-}, and $V_4O_9^{2-}$. These ions are stable under alkaline conditions.

Much of the analytical chemistry of vanadium is concerned with its use in ferrous and nonferrous metallurgy. Vanadium also finds application in catalysis and in the paint and ceramic industries. Environmental concerns about vanadium arise primarily from air-pollution problems. Vanadium can be released from fly ash and oil-combustion products. There are only a few references on vanadium speciation. One reference reported the simultaneous determination of V(IV) and V(V) [25]. Postcolumn reaction with PAR resulted in detection limits of about 10 ppb, even in the presence of high concentration of phosphate. Unfortunately, the studies were not carried out in samples. Urasa et al. [2] used DCPAE detection to speciate VO_2^+ and another vanadium species thought to be $VOCl_4^{2-}$.

11.5.7 Tin

Tin forms two stable inorganic species of Sn(II) and Sn(IV). Sn(II) is added to tin/lead alloy plating baths. The Sn(II)/Sn(IV) ratio is important to the plating bath performance. An IC separation was carried out with the use of 0.3 mM HCl eluent [26]. Neither Sn(II) or Sn(IV) were strongly retained by the cation-exchange column used in this work. Inorganic tin speciation is quite difficult, because Sn(II) hydrolyzes easily at neutral and alkaline pH.

Tin has a strong tendency to form organometallic complexes. Organotin compounds are used in marine antifoulant agents. Of course, there can be many tin species when combined with organics. MacCrehan [27] reported the separation of n-Bu_3Sn^+, Et_3Sn^+, and Me_3Sn^+ by cation chromatography and differential pulse amperometric detection. Organotin compounds tend to foul or poison the working electrode surface. The differential pulse technique used here eliminated this problem through the reoxidation of the reduction products on the electrode surface.

The speciation of tin in natural water was performed by Ebdon et al. [28] by AA detection. Sn(II), Sn(IV), and Bu_3Sn^+ were separated. An extraction/preconcentration sample preparation procedure was used. Jewett and Brinckman [29] used graphite furnace AA (GFAA) to detect several diorganotin and triorganotin species. Cation-exchange and reverse-phase chromatography was used in the several samples listed. Chromatographic separation of both dialkyltin and trialkyltin species appear to follow cation-exchange separation mechanisms.

11.5.8 Mercury

The harmful effects of mercury and organomercury species are well known. Disastrous effects, both on personal levels and large population levels, have resulted from exposure to certain mercury species. A method for Hg(I) and Hg(II) was reported which used on-column derivatization with diethyldithiocarbamate complexing agent. This process of using a complexing reagent seems appropriate for this speciation determination. In this way, the mixture can be "locked in" before the chromatography takes place. This is particularly important for metals that change oxidation states easily. A reverse-phase column and UV detection at 350 nm was employed [30]. Mercury compounds all appear to absorb UV light. The 254 nm detection has also been used with reasonable sensitivity and selectivity.

ICP-MS detection was used by Bushee [31] to speciate Hg(II) and thimerosol, a mercury-containing preservative. Other work showed the separation of $MeHg^+$, $EtHg^+$, and Hg(II). The detection limits were extremely good, at about the 1 ppb level.

2-Mercaptoethanol (ME), when added to an eluent, complexes with mercury compounds to produce charge-neutral compounds that can be separated on reverse-phase columns. MacCrehan et al. described this procedure in a separation of Hg^{2+}, $MeHg^+$, $EtHg^+$, and $PhHg^+$ [32,33].

11.5.9 Other Metals

Aluminum species exist in oxidation state III. In aqueous solution, the simple ion exists as $Al(H_2O)_6^{3+}$. This ion readily dissociates to give other ions such as $Al(H_2O)_5OH^{2+}$, all of which are colorless. Over a wide pH range, under physiological conditions, in alkaline solution, the species appear to be $Al(OH)_2^+$, $Al(OH)_3$, $Al(OH)_4^-$, $Al_3(OH)_{11}^{2-}$, $Al_6(OH)_{15}^{3+}$, and $Al_8(OH)_{22}^{2+}$. Study of the substitutions of aluminum aqua ion by ligands such as SO_4^{2-}, citrate, and EDTA has been established by ^{27}Al NMR spectroscopy [34].

A novel study on the speciation of aluminum in solution has been reported by Bertsch et al. [35]. Fluoro, oxalato, and citrate aluminum complexes were identified as distinct peaks together with free Al(III). Post-column reaction/UV detection was used. These studies were used in kinetic, ion-exchange, and toxicological investigations.

Gold cyanide complexes are important in gold-plating baths. As the Au(III) bath content increases, the plating efficiency is decreased. Mobile-phase ion chromatography can be used to determine total gold as well as $Au(CN)_2^-$ and $Au(CN)_4^-$ [35–37]. An anion-exchange method was also reported [39]. Conductivity or UV detection can be used.

Lead speciation of Et_3Pb^+ and Me_3Pb^+ along with some other organometallic species was reported by MacCrehan et al. [33]. Lichrosorb NH2 column and an ion-pairing type eluent was used. Bushee, Krull, and coworkers described an ion-pairing reverse-phase method for the separation of Cu(I) and Cu(II) [40]. Detection was by ICPAE. The effect on the retention of Cu(I) of changing the ion-pairing reagent from pentanesulfonic acid to octanesulfonic acid was shown. Mn(II) and Mn(III), along with some other metals, were separated on a C18 column and detected by W [41].

An elegant approach to determine Pt, Pd, and Au as their chloro complexes by IC was reported by Rocklin [42]. UV detection was used, resulting in detection limits of 0.03–1 ppm. In other work, IC was employed for the separation of chloro complexes of Pt, Pd, and Ir. UV detection was used with a 1 mM sulfosalicyclic acid, pH 4.2 eluent [43].

EDTA anionic complexes of Cd, Ni, Cu. and Zn have been separated by IC [44]. The results agreed well with those obtained by AA and ICP methods.

The separation and quantification of cyano complexes of various metals have been successfully carried out by Hilton and Haddad [45]. Cyano complexes of Cu(I), Ag(I), Fe(II), Fe(III), Co(II), Au(I), Au(III), Pd(II), and Pt(II) were analyzed by ion-pairing chromatography. The methods can be extended to many speciation problems.

References

[1]   W.T. Frankenberger, Jr., H.C. Mehra, and D.T. Gjerde, Environmental applications of ion chromatography, *J. Chromatogr.*, 504, 211, 1990.

[2]   I.T. Urasa, S.H. Ram and VD. Lewis, in Advances in Ion Chromatography, Vol 2,P. Jandik and R.M. Cassidy eds, Century International, Inc. Franklin, MA, 1990, p 93.

[3]   D.T. Gjerde, Eluent selection for the determination of cations in ion chromatography, *J. Chromatogr.*, 439, 49, 1988.

[4]   D. T. Gjerde, D. Wiederin, F.G. Smith, and B.M. Mattson, Metal speciation by means of microbore columns with direct-injection nebulization by inductively coupled plasma atomic emission spectroscopy, *J. Chromatogr.*, 640, 73, 1993.

[5]   D. Wiederin and J. Brennan, A low-pressure, valveless separation technique for ICP-MS, Cetac Technologies, Omaha, NE, unpublished results, 1999.

[6]   O. Shpigun and Yu.A. Zolotov, Ion Chromatography in Water Analysis, Wiley, New York, 1988.

[7]   Dionex Application Note TN24, May, 1987.

[8]   Cetac ANX Chromium Speciation Manual, Omaha, NE, 1999.

[9]   A. F. Geddes and J. G. Tarter, The ion chromatographic determination of chromium(III)-chromium(VI) using an EDTA eluant, *Anal. Lett.*, 21, 857, 1988.

[10] I.T. Urasa and S.H. Nam, Direct determination of chromium(III) and chromium(VI) with ion chromatography using direct current plasma emission as element-selective detector, *J. Chrom. Sci.*, 127, 30, 1989.

[11] M. J. Powell, D. W. Boomer, and D. R. Wiederin, Determination of chromium species in environmental samples using high-pressure liquid chromatography direct injection nebulization and inductively coupled plasma mass spectrometry, *Anal. Chem.*, 67, 2474, 1995.

[12] H. Saitoh and K. Oikawa, Simultaneous determination of iron(II) and -(III) by ion chromatography with post-column reaction, *J. Chromatogr.*, 329, 247, 1985.

[13] C.O. Moses, A.L. Herlihy, J.S. Herman and A.L. Mills, Ion chromatographic analysis of mixtures of ferrous and ferric iron, *Talanta*, 35, 15, 1988.

[14] J. Weiss, Handbook of Ion Chromatography, Dionex Corporation, Sunnyvale, CA, 1986.

[15] R.R. Turner, Oxidation state of arsenic in coal ash leachate, *Environ. Sci. Techn.*, 15, 1062, 1981.

[16] C.J. Craig, Organometallic Compounds in the Environment, Longman, London, 1986, p 198.

[17] C. McCrory-Joy, Single-column ion chromatography of weak inorganic acids for materials and process characterization, Anal. Chim. Acta, 181, 277, 1986.

[18] Personal communication, T. Matsuda, HP-Japan, 1999.

[19] Y.A. Zolotov, O.A. Shpigun, L.A. Bubchikova and E.A. Sedel'nikova, Ion chromatography as a method for the automatic determination of ions. Determination of selenium, Dokl. Akad. Nauk SSR, 263, 889, 1982, (Russian).

[20] S.G. Chen, K.L. Cheng and C.R. Vogt, Ion chromatographic separation of some aminopolycarboxylic acids and inorganic anions, *Mikrochim Acta*, 1, 473, 1983.

[21] R.J. Shamberger, Biochemistry of Selenium, Plenum, New York, 1873, p 185.

[22] D. Chakraborti, D.C.J. Hillman, K.J. Irgolic, and R.A. Zingaro, Hitachi Zeeman graphite furnace atomic absorption spectrometer as a selenium-specific detector for ion chromatography. Separation and determination of selenite and selenate, *J. Chromatogr.*, 249, 81, 1982.

[23] I.T. Urasa and F. Ferede, Use of direct-current plasma as an element selective detector for simultaneous ion chromatographic determination of arsenic(iii) and arsenic(v) in the presence of other common anions, *Anal. Chem.*, 59, 1563, 1987.

[24] H.C. Mehra and WT. Frankenberger, Jr., Simultaneous determination of selenate and selenite by single-column ion chromatography, *Chromatographia*, 25, 585, 1988.

[25] R.E. Smith, Ion Chromatography Applications, CRC Press, Boca Raton, FL 1988.

[26] R.A. Cochrane in Trace Metal Removal from Aqueous Solution, R. Thompson, Ed., The Royal Society of Chemistry, London, 1986, p 197.

[27] W.A. MacCrehan, Differential pulse detection in liquid-chromatography and its application to the measurement of organomethel cations, *Anal. Chem.*, 53, 74, 1981.

[28] L. Ebdon, S.J. Hill and P. Jones, Speciation of tin in natural waters using coupled high-performance liquid chromatography-flame atomic-absorption spectrometry, *Analyst*, 110, 515, 1985.

[29] F.E. Brinckman, W.R. Blair, K.S. Jewett and W.P. Iverson, Application of a liquid chromatograph coupled with a flameless atomic absorption detector for speciation of trace organometallic compounds, *J. Chrom. Sci.*, 15, 493, 1977.

[30] R.M. Smith, A.M. Butt and A. Thakur, Determination of lead, mercury, and cadmium by liquid chromatography using on-column derivatization with dithiocarbamates , *Analyst*, 110, 35, 1985.

[31] D.S. Bushee, Speciation of mercury using liquid chromatography with detection by inductively coupled plasma mass spectrometry, *Analyst*, 113, 1167, 1988.

[32] W.A. MacCrehan, R.A. Durst, Measurement of organomercury species in biological samples by liquid-chromatography with differential pulse electrochemical detection, *Anal. Chem.*, 50, 2108, 1979.

[33] V.l.A. MacCrehan, R.A. Durst and J.M. Bellama, Electrochemical detection in liquid chromatography: application to organometallic speciation, *Anal. Lett.*, 10, 1175, 1977.

[34] L.O. Ohman and S. Sjonberg, Equilibrium and structural studies of silicon(IV) and aluminum(III) in aqueous solution. Part 13. A potentiometric and aluminum-27 nuclear magnetic resonance study of speciation and equilibriums in the aluminum(III)-oxalic acid-hydroxide system, *J. Chem. Soc. Dalton Trans.*, 12, 2665, 1985.

[35] P.M. Bertsch and M.A. Anderson, Determination of gold, palladium, and platinum at the parts-per-billion level by ion chromatography, *Anal. Chem.*, 61, 535, 1989.

[36] Dionex Application Note 40R (1983).

[37] J. Weiss, Ion Chromatography, VCH, New York, NY 1995.

[38] Waters Chrom. Div. IC Series Application Brief No. 5001.

[39] M. Nonomura, Ion chromatographic analysis for cyanide, *Met. Fin.*, 85, 15, 1987.

[40] D. Bushee, I.S. Krull, R.N. Savage and S.B. Smith, Jr., Metal cation/anion speciation via paired-ion, reversed phase HPLC with refractive index and/or inductively coupled plasma emission spectroscopic detection methods, *J. Liq. Chromatogr.*, 5, 463, 1982.

[41]   B.W. Hoffman and G. Schwedt, Application of HPLC to inorganic analysis. Part VII. Comparison between pre-column- and on-column derivatization for separation of different metal oxinates; quantitative determination of manganese(II) besides manganese(III) ions, J. HRC and CC, 5, 439, 1982.

[42]   R.D. Rocklin, Determination of gold, palladium, and platinum at the parts-per-billion level by ion chromatography, *Anal. Chem.*, 56, 1959, 1984.

[43]   O.A. Shpigun and Yu. E. Pazuktina, Ion chromatographic determination of platinum in the presence of other platinum-group metals and inorganic anions , *Zh. Analit. Khim.*, 42, 1285, 1987.

[44]   M. Yamamoto, H. Yamamoto, Y. Yamamoto, S. Matsushita, N. Baba, and T. Ikushige, Simultaneous determination of inorganic anions and cations by ion chromatography with ethylenediaminetetraacetic acid as eluent, *Anal. Chem*, 56, 832, 1984.

[45]   P.R. Haddad and N.E. Rochester, Ion-interaction reversed-phase chromatographic method for the determination of gold(I) cyanide in mine process liquors using automated sample preconcentration, *J. Chromatogr.*, 439, 23, 1988.

12 Method Development

12.1 Introduction

When presented with an ion analysis problem, the worker may use a logical process of decision making to devise a working method for a new type of determination. The method may be based on one already published in the literature or on a standard method published by organizations such as the EPA, AOAC, or ASTM. Or the method may be entirely new, based on analyzing the problem of sample mixture and matrix.

There are probably several different methods that could be developed to solve the same problem. The decision to use one method over the other is frequently based on the availability of a certain column or detector. The method chosen may not even be the best available in a perfect world, but it may be the best given financial, time or instrumental constraints. Certain sacrifices may have to be made on sensitivity, accuracy, and the ease of analysis, e.g., the number of different runs needed. It is even possible a decision may have to be made on whether the total goal can be accomplished.

The chances of success are far greater if the literature is first searched. Good research papers or published methods are based on comprehensive testing of the proposed method. Gaining access to this experience can be quite valuable when faced with a new problem. Several papers may be published on a particular topic that, together, outline the various options available. Even a paper that shows limited success is useful because it can be compared to more successful methods.

12.2 Choosing the Method

12.2.1 Define the Problem Carefully

Exactly what is the analytical problem and what is the minimum analytical information needed to provide a reasonable answer? In this connection it is well to categorize the type of analysis desired: oxyhalides in drinking water, arsenic speciation in drinking water, speciation of chromium in plating baths, etc.

The second step is to describe the expected sample composition as completely as possible. For oxyhalide analysis, it is good to know which oxyhalide species are present, their estimated concentrations and what accuracy will be needed. The expected concentration of all cations and anions should be noted even if only one or the other is to be determined. Frequently, the counterion can form a complex with the sample ion of interest, thus making it less available for analysis. Also, any non-ionic materials such as alcohols, polymeric material, sugars or antioxidants should be listed as these may affect the analysis.

The next step is to consult the literature for information. Useful sources include books and standard methods from sources such as EPA, AOAC and ASTM, company literature, and journal articles. It is likely that other workers have faced situations similar or even identical to the problem at hand. It is interesting to note that difficult problems frequently garner more publications than easier problems. At this point, it should be possible to select an analytical method. The method selected is likely to reflect the personnel and equipment available and the worker's own preferences. The method selected need not be ideal; a method that simply works is sufficient. It is also necessary to consider any modifications needed to apply the selected method to your own analytical problem.

The final step is to confirm that the modified analytical method works for your particular situation. Does the resolution of peaks, sensitivity and reproducibility meet your requirements? Correct identification of each analytical peak is essential.

A standard for each unknown must be available and tried to determine if the method works. Normally, for method development, the sample is run without the standard and then a known amount of each standard is spiked into the sample. The peak that increases in height is identified as the ion of interest. The spike is normally obtained from concentrated standards so that volume of the sample does not change with the addition of the standard. For example, a spike of 100 μL of a standard into a 10 mL sample volume changes the volume by only 1 %. The concentration of the standard solution is chosen so that the increase in signal is measurable and appropriate. If a single concentration is spiked, then choosing a concentration that is expected to double the signal is appropriate.

12.2.2 Experimental Considerations

The different properties of anions and cations in the sample will affect the method development. These include whether the ion is organic or inorganic, multivalent or monovalent, and so on. The following discussion illustrates how these parameters can become important.

Anions can be conjugate bases of either strong acids or weak acids. Strong acid anions such as chloride or sulfate exist as the charged anion regardless of the eluent pH. Anions of weak bases such as acetate or formate are controlled by the eluent pH. If the separation mechanism is by ion exchange, then the pH must be high enough to ionize the anions and allow interaction with the column. Multivalent anions such as phosphate can change the extent of interaction by changing the pH. In many cases, a slight increase or decrease of eluent pH can result in improved resolution of a strong acid anion and weak acid anion pair. The easiest way to increase the eluent pH is to add (carbonate free) sodium or lithium hydroxide to the eluent. Lowering the eluent pH can be accomplished by increasing the concentration ratio of bicarbonate to carbonate. All other things being equal, changing the eluent concentration will affect divalent ions over monovalent ions. Increasing the concentration will shift the retention of divalent ions to shorter retention times faster than monovalent ions. The reverse is also true. Monovalent eluents are best for separating monovalent sample ions and divalent eluents are most useful for divalent sample ions. Certain commercial columns might have superior selectivity for monovalent anions and others are selective for divalent anions. Consult the column manufacturer for details of column selectivities.

Detection of weak acid anions is best by indirect conductivity detection or post-column reaction detection because the suppressed conductivity detection will not perform. Indirect conductivity detection is often used because the high pH used to separate the anions will also facilitate indirect conductivity detection of these anions. Chapter 4 describes a method of combining suppressed conductivity detection and nonsuppressed detection.

Conductivity detection is the most popular for ion chromatography. Although UV detection is often overlooked, it can be quite powerful. Amperometric detection, for example, offers selectivity and sensitivity, in many cases unsurpassed. The optimum eluent separation pH may not be the optimum pH for detection. An anion may be separated but not detected. This is especially true for some weak acid anions and suppressed conductivity detection. Chapter 4 discusses the use of different detectors for IC.

Anions that are polarizable will interact with the column to a stronger extent than other anions. Nitrate frequently will tail slightly because the anion will interact both with the column ion exchange site and the backbone of the ion-exchange matrix. Iodide, perchlorate, and many organic anions can be difficult to elute from the column in a sharp peak. Using a divalent driving eluent anions or addition of a small amount of organic solvent such as methanol may help. An eluent gradient also frequently helps elute a range of anion types. Depending on other materials also to be analyzed, polarizable ions may be easier to separate by ion-pairing chromatography. Ion-exchange separations are more resistant to changes in sample matrix. In ion-pairing separations, the sample matrix can cause the sample retention to shift to different times. This is less likely to happen in ion-exchange chromatography.

Several charts of retention times for specific columns and eluents are listed in different chapters in this book and in company literature. While new columns are introduced, these charts can still be used as tools to determine the relationship between ions. Usually a combination of the table and a chromatogram will help predict what a chromatogram should look like. Keep in mind that the weak acids are affected most by eluent pH. Divalent ions are affected most by eluent concentration.

Ion exclusion is useful for the separation of weak acid anions. The decision to use ion exclusion over ion exchange frequently depends on the matrix of the sample. If a mixture of weak acid anion and strong acids anions is to be analyzed, then ion exchange is a separation tool that will be the most effective. However, if weak acid anions are to be analyzed in the presence of large concentrations of strong acid anions, e.g., the determination of acetate in hydrochloric acid, then ion exclusion will be the superior separation tool. Chloride, sulfate and similar anions are not retained by ion exclusion and therefore high matrix concentrations of these anions can be tolerated.

Sample preparation (Chapter 9) includes simple procedures such as centrifugation or filtration. Other more complex sample preparation procedures include passive or active dialysis, preconcentration, combustion or precipitation of matrix ions. In some cases, choosing the correct eluent/column/detector system will negate the need for sample preparations. A selective detector will be able to detect a minor analyte in the presence of other ions. Ion exclusion allows the passage of strong acid anions prior to separation of weak acid anions. Methods that require the least sample preparation or minimal sample preparation directly prior to injection are the most desirable.

12.3 Example of Method Development

12.3.1 Examining the Literature and the Problem

In order to illustrate the method development process, a description of a problem is presented. For example, one might want to determine sulfite in wine. Sulfite is a widely used food preservative and whitening agent. There have been many reported cases of allergic reaction to the ingestion of sulfite contained in foods or beverages. Since 1986, the FDA has required warning labels on any food or beverage containing more than 10 mg/kg or 10 mg/L of sulfite, respectively.

The concentration of sulfite in wine is expected to be in the low ppm range; however, the sample matrix includes a wide variety of materials including several carboxylic acids, ethanol, sugars, and many other components. A search of the literature for sulfite determinations by ion chromatography yielded several references [1–22] describing a variety of separation methods, matrixes, and detection methods. Detection methods included ICP-AE [1], amperometric [3,5,9] refractive index [17], fluorometric [8,14] UV [12] and other methods. The sample type ranged from vitamins [8,14] to photographic fixers [13], animal feed [8,14], and food [10,18]. The search also found a paper on capillary electrophoresis [4]. Vendors including Dionex [19], Zellweger [20], Metrohm [21], and Sarasep [22] have published separations of sulfite.

The best reference that was found was AOAC method 990.31 [18] that was developed by Kim and Kim. The method was based on extensive work with a variety of foods and beverages. The incentive for the work was to find an alternative to the modified Monier-Williams method [23] which is time-consuming and labor-intensive.

The Kim and Kim method uses ion-exclusion chromatography with a dilute sulfuric acid eluent. Detection is based on amperometric detection with a Pt working electrode. The method is quite selective. Samples are only blended with a high pH solvent to extract both bound and unbound sulfite, and then filtered and injected into the ion chromatograph. Mannitol is added to the extraction solvent to slow the oxidation of sulfite to sulfate.

After examining the rest of the literature, several observations can be made. First, the most sensitive detection method is amperometric detection with a Pt working electrode. Other detectors may work, but may not be practical.

ICP-AE or fluorescence work well but may not be available to the user.

Ion exclusion is the most preferred separation for sulfite determinations. However, considering all of the neutral materials that are contained in wines (sugars, organic acids, and alcohols), ion exchange may be considered. In many anion-exchange columns, sulfite and sulfate will coelute and this will be a factor if conductivity or other type of general detection method is used. Information is available from column vendors on which columns will resolve sulfite and sulfate. It should be noted that the eluent pH will determine whether sulfite is an anion or neutral. This is important to consider when choosing between ion exclusion and ion exchange. Also, the optimum eluent pH may not be the optimum detection pH.

12.3.2 Conclusions

The Kim and Kim ion exclusion method works well. The amperometric detection procedures used are extremely selective and sensitive. However, this detection method is not without problems. Although good results have been reported, for some cases, the sample matrix material can foul the electrode leading to non-reproducible peak areas and/or lost of sensitivity. Pulsed amperometric detection (see discussion in Chapter 4) has been suggested as an improved detector for this method [19] in order to prevent deactivation of the working electrode. Modern DC amperometric detectors are computer-controlled and oxidation cleaning and reducing potentials can be applied to the working electrode after the run is completed. So a PAD detector may not be necessary as long as a routine cleaning operation is implemented into the analysis.

Sample preparation can be quite important in sulfite determination since sulfite can easily be oxidized to sulfate. So in this case, stabilization or storage of the sample is also considered to be sample preparation. For two of the references discussed, aldehyde adducts are to be formed to stabilize the sample [2,9]. Other work described a sulfite–disulfite equilibrium process [7]. It is important that the eluent does not change the sample during the elution process. Eluents do not normally contain oxidizing agents except for, perhaps, dissolved oxygen. Degassing the eluent will remove dissolved oxygen (Chapter 1).

The optimum sample preparation appears to be keeping the sample tightly capped and out of sunlight until just before the injection. The sample should be quickly removed and analyzed. In general, it is better not to perform extensive sample preparation unless necessary. Of course, in many cases, sample preparation is needed. Although adding manitol to the sample extraction solvent is not essential, it is probably important to ensure good results.

This is just one example how a method may be developed. There are other examples. The best method determination of trace anions in concentrated acids either requires removing the matrix anion, selective detection, or choosing a column that has sufficient capacity and selectivity to allow the matrix to travel quickly through the column while the analytes are retained and separated. A good method for determining chloride in a 1 % solution of boric acid solution is to use non-suppressed ion chromatography with conductivity detection. A eluent is chosen with a low pH (pH 4.5 phthalate, for example) so that the boric acid passes quickly through the column and chloride is retained and detected. Using a high-pH eluent would swamp the column because the borate would be ionized and retained.

Developing a method requires careful thought, an open mind, and just plain hard work evaluating the literature or trying out procedures. This book has tried to present fundamental principles of ion chromatography with the hope that using this information will make IC methods more productive.

References

[1]   D. R. Migneault, Enhanced detection of sulfite by inductively coupled plasma atomic emission-spectroscopy with high-performance liquid-chromatography, *Anal. Chem.*, 61, 272, 1989.
[2]   T. Sunden, M. Lindgren, A. Cedergren, D. D. Siemer, Sulfite stabilizer in ion chromatography, *J. Chromatogr.*, 663, 255, 1994.
[3]   R. Leubolt, H. Klien, Determination of sulfite and ascorbic acid by high-performance liquid chromatography with electrochemical detection. *J. Chromatogr.*, 640, 271, 1993.
[4]   D. R. Salomon, J. Romano, Applications of capillary ion electrophoresis in the pulp and paper industry, *J. Chromatogr.*, 602, 219, 1992.
[5]   H. R. Wagner, and M. J. McGarrity, The use of pulsed amperometry combined with ion-exclusion chromatography for the simultaneous analysis of ascorbic acid and sulfite, *J. Chromatogr.*, 546, 119, 1991.
[6]   J. E. Parkin, High-performance liquid chromatographic investigation of the interaction of phenylmercuric nitrate and sodium metabisulfite in eye drop formulations, *J. Chromatogr.*, 511, 233, 1990.
[7]   B. J. Johnson, Sulfite-disulfite equilibrium on an ion chromatography column, *J. Chromatogr.*, 508, 271, 1990.
[8]   S. M. Billedeau, Fluorimetric determination of vitamin K3 (menadione sodium bisulfite) in synthetic animal feed by high-performance liquid chromatography using a post-column zinc reducer, *J. Chromatogr.*, 471, 371, 1989.
[9]   J. F. Lawrence, and F. C. Charbonneau, Separation of sulfite adducts by ion chromatography with oxidative amperometric detection, *J. Chromatogr.*, 403, 379, 1987.
[10] J. F. Lawrence and K. R. Chadha, Headspace liquid chromatographic technique for the determination of sulfite in food., *J. Chromatogr.*, 398, 355, 1987
[11] N. Sadlej-Sosnowska, D. Blitek and I. Wilczynska-Wojtulewicz, Determination of menadione sodium hydrogen sulfite and nicotinamide in multivitamin formulations by high-performance liquid chromatography, *J. Chromatogr.*, 357, 227, 1987.
[12] R. G. Gerritse and J. A. Adeney, Rapid determination in water of chloride, sulfate, sulfite, selenite, selenate, and arsenate among other inorganic and organic solutes by ion chromatography with UV detection below 195 nm, *J. Chromatogr.*, 347, 419, 1985.
[13] J. M. McCornick and L. M. Dixon, Determination of sulfite in fixers and photographic effluents by ion chromatography, *J. Chromatogr.*, 322, 478, 1985.
[14] A. J. Speek, J. Schrijver and W. Schreurs, Fluorimetric determination of menadione sodium bisulfite (vitamin K3) in animal feed and premixes by high-performance liquid chromatography with post-column derivatization, *J. Chromatogr.*, 301, 441, 1984.
[15] A. Gooijer, P. R. Markies, J. J. Donkerbroek, N. H. Welthorst, and R. W. Frei, Quenched phosphorescence as a detection method in ion chromatography: the determination of nitrite and sulfite, *J. Chromatogr.*, 289, 347, 1984.
[16] W. E. Barber and P. W. Carr, Ultraviolet visualization of inorganic ions by reversed-phase ion-interaction chromatography, *J. Chromatogr.*, 260, 89, 1983.
[17] P. R. Haddad and A. L. Heckenberg, High-performance liquid chromatography of inorganic and organic ions using low-capacity ion-exchange columns with indirect refractive index detection, *J. Chromatogr.*, 252, 177, 1982.

[18] AOAC Official Method 990.31 in Official Methods of Analysis of AOAC International, 16th ed., Vol. II, P. Cunniff, ed, 1995.

[19] Dionex Application Note 54, Determination of sulfite in food and beverages by ion exclusion chromatography with pulsed amperometric detection, Dionex Corp., Sunnyvale, CA, 1999.

[20] Lachat Ion Chromatography Data Pack, Zellweger Analytics, Milwaukee, WI, 1999.

[21] Metrohm Application Note S-12, Determination of lactate, chloride, nitrate, sulfite and phosphate in wine, Metrohm Ltd, Herisau, Switzerland, 1999.

[22] Sarasep Application Note, Determination of sulfite by ion exchange and conductivity detection, San Jose, CA 1996.

[23] AOAC Official Method 962.16 in Official Methods of Analysis of AOAC International, 16th ed., Vol II, P. Cunniff, ed., 1995.

Index